BANG!

THE ULTIMATE BOOK OF EXPLOSIONS BIG AND SMALL

ROD GREEN

BANG!

THE ULTIMATE BOOK OF EXPLOSIONS BIG AND SMALL

BARNES
&NOBLE
BOOKS
NEW YORK

CONTENTS

Introduction	6
The Big Bang	10
A to Z of Bangs	18
The Bang that Shook the World!	22
Bang to Light	24
The Gods Are Angry!	32
Krakatoa	34
Bang Goes the Engine	36
Blowing You Away	40
The Deadly Custard Bomb	48
Why Balloons Go Bang!	52
El Alamein	56
Danger—Explosives!	58
Bangs in Space	64
Drum Beating	68
Feeling the Heat	70
Bottled Bangs	76
The *Mont Blanc* Tragedy	78
Rockets	80
Jet Engines	86
Listen Up!	90
You're Fired!	94
Smashing the Sound Barrier	98
Underwater Bangs	100
Volcanoes	102
Nature's Arsenal	106
A Case of Gas	112
Bombs Away!	116
Shake, Rattle 'n' Roll	122
The *Hindenburg*	126
Power Station in the Sky	128
Atomic Explosions	132
Watch the Birdie	138
Icicle Pops	140
A Night at the Theater	142
Hot Stuff	144
Lethal Bangs	148
Fireworks	152
Index	156
Picture Credits and Acknowledgments	160

BANG!

"**WHAT THE HECK WAS THAT?**" might well have been your reaction if you could have heard the word above instead of just reading it. Try reading it aloud. "BANG!" Now try reading it *really* loud. "*BANG!*" If you're reading this on a bus or in the subway, the people around you will be getting pretty nervous by now. Never mind, it might buy you a bit of elbow room!

So what was it? It was you, you fool. It was you vibrating your vocal chords to produce a sound. The sound started off way down in your throat and was then amplified in the echo chamber of your mouth where you fine-tuned it, pursing your lips and releasing it in a burst to produce a noise that sounds a bit like an explosion. In many ways, it was an explosion, or part of one—almost all bangs are. In fact, if you take a bang to be what you hear, almost all bangs are exactly the same.

All the same? A bursting balloon is the same as a cannon firing? No, not entirely the same but to understand the difference you need a degree in chemistry, a

BONK!

doctorate in applied physics and a professorship in quantum mechanics… or a copy of this book. Yes, this is your key to understanding all things around you that go BANG! and quite a few things that go BONK! and some stuff that goes THUD! but nothing that goes PLOP! We had to draw the line somewhere!

Throughout the book we will look at bangs of all shapes and sizes from the Big Bang that started it all, way back at the

beginning of the universe, through to the kinds of bangs we can hear every day: common sights and sounds in our streets like a backfiring car, music playing or a thunderstorm.

Man versus Mother Nature

Prior to August 1945, kites wheeling in the sky above Nagasaki were a common sight. These large, scavenging birds were among the very first victims of the second atomic bomb dropped on Japan. One survivor of the blast remembers seeing some of the birds walking around on the ground on the outskirts of the town, completely bald.

All their feathers had been burned off by the explosion while they were in mid-flight. Appalling as this sounds, and it is true, it is one of the least of the horrors unleashed upon the world by man's invention of the atomic bomb. The nuclear bang may be the biggest bang man has yet produced, but compared with Mother Nature, man is a complete amateur. Still, man has only been at it for 100,000 years or so while Mother Nature has had 15 billion years to play around with bangs, and we should really hope that man never gets to be as good at it as she is. Even the lesser, non-nuclear bangs man has mastered are quite scary enough.

There are plenty of non-scary bangs to be examined, though. By non-scary, of course, I mean they are relatively harmless, not that they won't make you jump out of your skin! We have to have the scary bangs in the book, and they inevitably bring with them sad situations and tragic loss of life, but we can also talk about nice bangs that everyone loves—the popping of a champagne cork, the bursting of fireworks, the drumbeat of dance music and the warmth of the sun are all bangs that bring smiles to our faces. These are all bangs to be enjoyed and they all feature in this book. Hang on—the warmth of the sun is a bang? Oh, yes... the most important bang you'll never hear. All will be revealed within these pages.

Don't worry if science has never been your strongest subject. You should be able to follow what's going on in this book, even if you can't tell your periodic table from your breakfast table! You might have a problem working out the atomic number for maple syrup but don't let that put you off. Try marmalade instead!

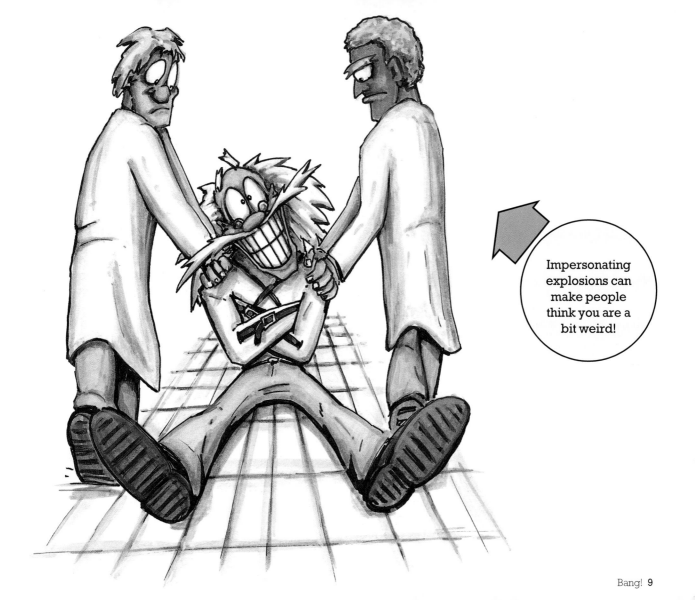

The sun

OUTER TEMPERATURE
9932°F (5500°C)

INNER TEMPERATURE
27 000 000°F (15 000 000°C)

THE BIG BANG

THE FIRST MORNING IT WAS PRETTY HOT. How hot? Well, on a hot summer's day the temperature might be about 86°F (30°C), but it was a lot hotter than that. In one of the hottest places on earth, Death Valley in California, the temperature can reach a stifling 120°F (49°C), but Death Valley was cool by comparison. At the earth's core, a molten fiery mass, the temperature is estimated to be between 5400°F and 7200°F (3000—4000°C), but that doesn't even come close. The surface of the sun is thought to be about 9900°F (5500°C). It doesn't come much hotter than that, but it did on that first morning.

In fact, on the very first morning, you'd have known the day was going to turn out to be a real scorcher! Within the first immeasurably small fraction of a second the temperature soared to trillions of degrees. This was the Big Bang.

The beginning of everything

Of course, this can't really be described as the first morning. We think of "morning" as being when the sun rises, but at this

Death Valley, California, where the temperature can reach a stifling 120°F (49°C).

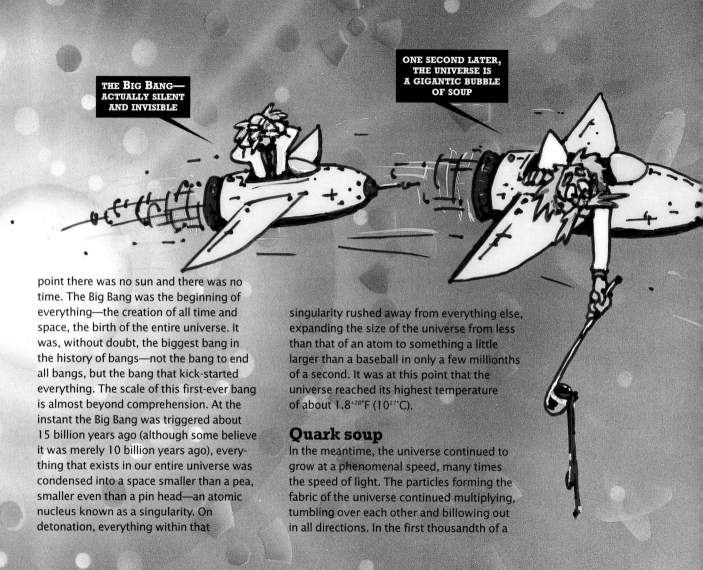

THE BIG BANG—
ACTUALLY SILENT
AND INVISIBLE

ONE SECOND LATER,
THE UNIVERSE IS
A GIGANTIC BUBBLE
OF SOUP

point there was no sun and there was no time. The Big Bang was the beginning of everything—the creation of all time and space, the birth of the entire universe. It was, without doubt, the biggest bang in the history of bangs—not the bang to end all bangs, but the bang that kick-started everything. The scale of this first-ever bang is almost beyond comprehension. At the instant the Big Bang was triggered about 15 billion years ago (although some believe it was merely 10 billion years ago), everything that exists in our entire universe was condensed into a space smaller than a pea, smaller even than a pin head—an atomic nucleus known as a singularity. On detonation, everything within that

singularity rushed away from everything else, expanding the size of the universe from less than that of an atom to something a little larger than a baseball in only a few millionths of a second. It was at this point that the universe reached its highest temperature of about 1.8^{+28}°F (10^{27}°C).

Quark soup

In the meantime, the universe continued to grow at a phenomenal speed, many times the speed of light. The particles forming the fabric of the universe continued multiplying, tumbling over each other and billowing out in all directions. In the first thousandth of a

AFTER A FEW SECONDS, THERE'S A MAJOR PROBLEM WITH GAS!

FIFTEEN BILLION YEARS LATER, THE UNIVERSE IS AS WE KNOW IT TODAY

second, the universe doubled in size over a hundred times to become over ½ mile (0.8 km) wide. This was a ½-mile (0.8-km) pool of super-heated plasma "soup" with sub-atomic particles called quarks forming, bouncing off each other, destroying each other, reforming and all the time hurtling outward from the point of the original singularity and away from each other. By the time the universe was one second old, the temperature had cooled to only a few billion degrees. Then, some rather interesting changes began to take place. The "soup" began to thicken, or at least turn a bit lumpy, as the quarks clung together to form protons and neutrons. After about three minutes, the protons and neutrons began reacting with each other to create the basis for hydrogen and helium atoms and the temperature dropped to less than 1.8 billion degrees Fahrenheit (1 billion degrees centigrade).

Let there be light!

After all the excitement of those first few nanoseconds, things started to calm down somewhat and the various elements that were forming did a bit of mixing and swirling for 300,000 years or so until they began to create gases. This is when another phenomenon probably started to appear—light. Until then, despite the fact that the biggest bang of all time had been set off, there had been no flash, no bright flame, not even a flicker. Light did not exist until the nuclear reactions that were going on reached the stage where they actually produced light; the continued expansion of all things also served to allow different kinds of energy to separate. Light was then able to escape from matter. Given that all of this was going on in the vacuum of space one must assume that when the Big Bang detonated, it did so silently. Sound, after all, cannot travel through a vacuum. "In space," as the famous movie trailer pointed out, "no one can hear you scream." So, had there been anyone around to listen, the biggest bang ever known would have been completely inaudible and invisible. So, the Big Bang was more of a Big Black Silence, really.

Gases galore

Nevertheless, 300,000 years after the initial explosion there was still plenty going on, although it might not have appeared all that obvious, and by now there was some illumination, too. The great attraction, the show not to be missed over the next 700,000 years, was the formation of gigantic swirling clouds of gases such as hydrogen and helium. You would need to be wrapped up pretty warmly if you were spectating from an open grandstand, though, because it was now becoming distinctly chilly. The overall temperature in the universe had dropped to about -328°F (-200°C). There were hot spots, of course. Inside those swirling gas clouds, gravity was taking effect and galaxies were starting to form. Under certain circumstances, the gases formed clumps within the clouds. These clumps grew more and more dense until, finally, they collapsed in on themselves to form the first stars, which were just massive bubbles of burning gas.

The universe as we know it

Creation was now in full swing, but there was to be no settled order, no peaceful evolution. Billions of years on, the Big Bang began working like a giant firework, the initial aerial explosion spawning an infinite number of secondary pyrotechnic displays. Galaxies started to cluster together as gravity really began to take hold and the earliest stars started to behave like foundries running out of control. In their death throws, they hurled out heavy elements created in the furnaces of their interiors. These elements would eventually combine, cool and transform themselves into other stars and planets. After

EDWIN HUBBLE 1889–1953

BORN IN MISSOURI IN 1889, Edwin Hubble was more interested in sports than science as a youngster. He won a Rhodes Scholarship to study law at Oxford, and started taking astronomy seriously on his return to the U.S.A.

Hubble began theorizing on the existence of different galaxies in the universe as island-like clusters of stars and planets in space. In 1924 he calculated the distance to the Andromeda nebula to be a hundred thousand times further than the distance to our nearest stars. In studying the distances between the galaxies, Hubble began to develop another theory. He was soon able to show that, because of the way the light emitted by distant galaxies changes when seen from Earth, the changes can be measured to show the galaxies are drifting further apart. It was then logical to assume that at one point, everything in the universe was far more condensed and the thing that had started the whole movement must have been a pretty Big Bang.

Hubble died in 1953 before man's great leaps into space and before much of the evidence supporting his Big Bang theory had emerged. In June 1995, for example, scientists first detected primordial helium in the outermost areas of the universe, as would have been expected with the Big Bang, since hydrogen and helium would have been created at the beginning of everything. Hubble, of course, did not have the sophisticated technology available today when he first formulated his theory. His observations were made using what was, at that time, the most advanced astronomical equipment in the world, the 100-inch (2.5-m) telescope at Mount Wilson in Southern California. Scientists who continue to explore Hubble's theories today use a telescope bearing the great man's name—the Hubble Space Telescope.

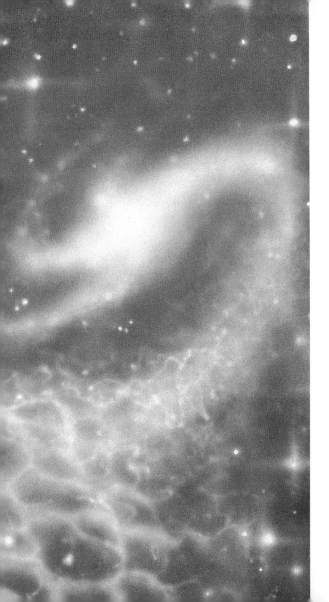

15 billion years, the "outside" temperature was -454°F (-270°C) and the universe was as we know it today.

It's easy to get carried away with all this talk of millions, billions, trillions and creation on a grand scale, but how do we know that any of this ever actually happened? Who on earth came up with this bizarre idea in the first place? Edwin Hubble, that's who (see page 15).

The debate continues

Despite evidence that has continued to emerge supporting Hubble's Big Bang theory, not everyone believes it to be true. Some scientists believe that the changes in light from distant galaxies can be explained by other phenomenon, challenging the assertion that the galaxies are drifting apart and throwing the whole debate wide open. Nevertheless, the theory persists as a viable attempt to explain not only the creation of the stars, the planets, the geography of space and the age of the universe, but also the creation of the first atoms, the first elements, the first substances to exist.

If you accept that the Big Bang happened, however, and that the universe is still expanding, then a number of other perplexing questions pop up. If this Big (Silent, Invisible) Bang did happen 10 or 15 billion years ago when the building blocks for the entire universe were crammed into that "singularity" less than the size of an atom, where was that singularity? Into what did the embryonic universe expand? If it grew to become ½ mile (0.8 km) wide, what was outside its ½-mile (0.8-km) perimeter? If the universe is still growing, what now exists beyond its ever-expanding edge? Just where the heck are we and what are we doing here? And who set off that Big Bang in the first place?

If you persist in thinking along these lines, then the next Big Bang will probably come when your brain explodes!

IF YOU THINK ABOUT THE UNIVERSE TOO HARD, THEN THE NEXT BIG **BANG** WILL PROBABLY COME WHEN YOUR BRAIN **EXPLODES**!

The temperature at the inception of the Big Bang is thought to have been around 1.8^{+28}°F (10^{27}°C). 10^2 is 100 and 10^3 is 1,000, so 10^{27} is … rather a lot!

Sound won't travel through a vacuum, so you can't hear an explosion in space.

The Big Bang theory was invented by Edwin Hubble from Missouri.

Hydrogen and helium are two of the oldest elements in the universe. Helium is inert but hydrogen is highly explosive.

A TO Z OF BANGS

A Atomic bomb—the first atomic bomb was detonated in a test in the desert of New Mexico in 1945. Although only two nuclear devices have been used in anger—those dropped on Japan at the end of World War II—there have been hundreds of test detonations, with 178 taking place in 1962 alone.

B Big Bang—our whole universe started with a bang (the biggest bang ever) around 15 billion years ago and the party just keeps rolling on...

C C4—Composition 4, containing the explosive cyclotrimethylenetrinitramine, is one of the most powerful modern plastic explosives. Also known as RDX, which stands for "research development explosive," it is popular with both military and terrorist organizations around the globe.

D Dynamite—invented in the mid-19th century by Swedish chemist Alfred Nobel, dynamite is an explosive, which includes the highly unstable nitroglycerin. Nobel's dynamite compound made nitro a little safer to handle and therefore more user friendly.

E Earthquakes—some of the world's biggest natural bangs are produced when the plates that make up the earth's crust bash together. The power released can literally cause the ground to ripple like the surface of a pond.

F Fireworks—invented by the Chinese a thousand years ago to make loud bangs, drive away evil spirits from religious or ceremonial events and scare the pants off the old lady next door!

G Gunpowder—the basic recipe is believed to have been discovered by the Chinese and brought to Europe by explorer Marco Polo, although the Indians and the Arabs would almost certainly disagree with this Eurocentric historical perspective.

H *Hindenburg*—the biggest airship ever built, the *Hindenburg* was a balloon full of another "H"—hydrogen. How safe did anyone ever think an 800-foot (245-m) long, cigar-shaped bag of explosive gas would be? Not very, as it turned out...

I Infra-red—the sort of bangs produced by stars as they go about their day, or nuclear bombs as they ruin someone's day, emit electromagnetic radiation of which infra-red rays are a part.

J Jor-el—father of Superman, Jor-el launched his son into space in a capsule destined for Earth when he discovered that their home planet, Krypton, was about to go bang.

K Krypton—Superman's home planet, which went bang.

L Light—when a major explosion is under way, the disturbed electrons of agitated atoms release packages of energy called photons, which create what we know as light.

M *Mont Blanc*—one of the biggest-ever accidental explosions happened in 1917 during World War I when a munitions ship called the *Mont Blanc*, thousands of miles from the nearest battlefield, caught fire in Halifax Harbor, Nova Scotia.

N Nitroglycerin—the first high explosive, invented by Italian Ascanio Sobrero in 1846. This bizarre compound had a freezing point of about 55°F (13°C), so it would turn to liquid on a mildly warm day and could detonate simply from the impact of being dropped on the ground.

O Oxygen—any explosion is really a kind of fast-burning fire and doesn't get far without oxygen to sustain it. But then, who does?

P Pyrotechnics—officially the art of staging a jaw-droppingly spectacular fireworks display and therefore one of the best jobs in the world... unless it rains!

Q Quarks—believed to have been produced by the Big Bang at the beginning of time and to have bonded together in small bunches to produce the constituent particles that go to make up the different parts of an atom.

R Rockets—not only do many of them deliver some of the mightiest explosive payloads ever squashed into a nose cone, rockets are powered by a continuous bang. A controlled explosion expands gases, which push out of the rocket's rear end to drive it forward.

S Sun—nuclear explosions, are they really all bad? Not if they are happening on our sun. Without them we would all be freezing in the dark.

T TNT—the three letters you see stenciled alongside "ACME" on almost every package received by Wile E Coyote and destined for use against his adversary, pesky Road Runner. TNT is trinitrotoluene, a high explosive. If you want to know what ACME is, go ask the coyote.

U Ultraviolet—like X-rays, gamma rays and other electromagnetic phenomenon produced by, among other things, nuclear explosions, ultraviolet radiation is a type of energy wave. A little exposure to a fairly weak source will give you a nice tan. A lot of exposure will fry you to a crisp.

"V" is also for volcano. The world's biggest bangs have come from volcanic eruptions

V V-1 and V-2—nasty Nazi flying bombs used in World War II, which packed a powerful punch. The V-1 was an early type of jet-powered cruise missile while the V-2 was a supersonic rocket.

W Wernher von Braun—the mainstay of the Nazi rocket program during World War II. Von Braun's team developed the V-2 before being whisked off to America after the war to play a key role in the U.S. space program.

X X-rays—like ultraviolet rays, X-rays are a form of radiation energy wave. X-rays can pass through healthy human flesh and tissue, but will not penetrate other obstructions in the body so well. They are therefore used to produce "shadow" pictures of our insides.

Y Yaeger—American General Chuck Yaeger was a World War II fighter ace who evaded capture after being shot down over enemy territory. He became the first man officially to fly faster than the speed of sound and create a sonic boom.

Z Z particles—estimated to be a hundred times the mass of a proton, Z particles, along with W particles, facilitate the interactions that cause nuclear decay.

BANG

FAMOUS BANGS

THE BANG THAT SHOOK THE WORLD

THE METEORITE HURTLED TOWARDS **E**ARTH at a speed of over 1,000 miles (1,600 km) per minute. There was to be no last-second swerve, no miraculous reprieve—it was going to hit. The impact would be in the area of Sudbury in Ontario, Canada, and at over 5½ miles (9 km) wide, this space rock was going to make quite a dent! It was going to be one of the biggest bangs ever, a real window rattler—or it would have been, had there been any windows to rattle. Over 160,000 people live in Sudbury today but at the time of the meteorite strike the population was zero. The impact, you see, happened 1.87 billion years ago.

Making an impact

There was no one around to witness the event, no one to launch nuclear missiles at the approaching meteorite to try to divert it, and no one to land a space shuttle on it to try to blow it up, Hollywood-style. That's rather fortunate really, because if a chunk of space debris this size collided with Earth today, it wouldn't only be the people of Sudbury who would suffer, it could easily wipe out all life on the entire planet.

The force of the impact of a rock this size—about 35 cubic miles (over 140 cubic km)—smashing into the ground at a speed approaching a hundred times the speed of sound, would have the same effect as the detonation of several billion tons of TNT. It would make the Hiroshima atomic bomb look like a cheap firework. It would immediately wipe out everything within a radius of about 500 miles (800 km).

Earth tremors from the strike would be felt all over Canada and the United States and would send seismographs wild all over the world.

Catastrophic consequences

The Sudbury impact formed a crater 8 miles (13 km) deep and up to 150 miles (240 km) in diameter, hurling molten material, rocks and other debris into the air with such force that some of it would have broken free from the earth's atmosphere and gravity to head off into space, perhaps to cause similar craters on the moon or to impact on some other far distant planet. The "fallout" cloud from the impact would have spread far and wide, circling the earth. For the modern world, this would be a complete catastrophe.

If this impact happened today, the cloud of dust and debris would block out the sun's rays, causing a rapid cooling across the whole world. This "impact winter" and the prolonged darkness would cause plant life to

die off and, consequently, animals that feed on plants to die off. Arctic conditions would spread from the Poles in a new ice age and even though areas nearest the equator might avoid the ice, bereft of sunlight they would experience the end of all plant and animal life. That would probably mean the end of all human life, too—complete extinction.

Could it happen?

It certainly could. It is estimated that a major impact happens every 50 million years or so and there have been several comparable with the Sudbury meteorite. Many scientists believe that an impact at Chicxulub in the Yucatan, Mexico, was responsible for the extinction of the dinosaurs. Many also believe that we are overdue for another big bang just like it and governments around the world are now taking the threat seriously, establishing projects to monitor NEOs (near earth objects).

While monitoring might help give us prior warning of a space rock heading for Earth, there is, as yet, no viable way of stopping something like the Sudbury meteorite.

BANG

The first atomic bomb was detonated in the desert in New Mexico, U.S.A., in July 1945

BANG TO LIGHT

IN JULY 1945, Lieutenant Joe Wills was driving his sister-in-law, 18-year-old Georgia Green, to the University of New Mexico where she studied music. It was 5.30 a.m. and suddenly the pre-dawn sky was lit by a bright flash of light. Georgia gripped her brother-in-law's arm and asked "What's that?" Although Georgia had been blind since early childhood, she still "saw" the flash.

The first atomic bomb

The flash was caused by the detonation of the world's first atomic bomb at Trinity in the New Mexico desert, about 50 miles (80 km) from where Georgia and Joe were driving along Highway 85. The flash was the first they knew that anything out of the ordinary had happened but it was so incredibly bright that, even though they were so far from the test site, Georgia, who was only marginally sensitive to the difference between darkness and light, immediately recognized that something had happened.

This was, of course, the first really big man-made bang, but why did Georgia see the flash rather than feel a blast or a heat wave? And why does a really big bang create so much light anyway?

Seeing the light

To shed some light on the light conundrum we must first look at how explosions work and how light works.

An explosion is basically a fire, but a fire that happens very quickly. All kinds of fire, from the camp fire that cowboys warm their boots on to the raging infernos that ravage entire buildings, are part of a chemical reaction. The chemical reaction that takes place is matter (be it the wood of the camp fire or the fabric of a building) changing its form, the atoms that make up that matter being agitated by heat, friction or some other input of energy to the extent that they break apart, combine with oxygen and form gases such as carbon dioxide or hydrogen compounds, which in turn form part of what we recognize as smoke.

In the process of becoming agitated and energized, the atoms of the matter expand. Imagine that the electrons surrounding an atom orbit the atom like a satellite orbiting the earth. When some form of energy is applied to the atom, the electrons absorb the energy by using it to orbit at a greater distance, like a satellite moving further away from the earth. When the electrons drop back to their normal orbiting "altitude," they release a burst of energy called a photon. These photons are different colors depending on how much energy the atom's electron releases. The higher an orbit it achieves, the more energy it releases when it drops back again.

Multiply this effect by a number of atoms bigger than you can count on all of your fingers until the day you die, and you have a stream of energy radiating out from the original source. And traveling quite quickly. Light zips along at about 186,000 miles (299,350 km) per second. How fast is that? Well, if the speed limit in your street is 30 mph (48 kmph), then light is looking at a whole book of speeding tickets! You can pick up a pretty hefty fine for doing twice the legal speed limit. Imagine how long it would take to pay off your ticket if you were caught doing the speed of light, at over 22 million times the legal limit!

This is why Georgia and Joe saw the flash first rather than observed any of the other phenomena created by the atomic explosion, such as hearing the bang.

All the colors of the rainbow are actually all the colors of white light, refracted through water droplets in the air

Traveling at the speed of light

Light travels faster than sound. Light, in fact, travels faster than anything else we know. Nothing exceeds the speed of light. Traveling at that speed of course, is utterly reckless. A major accident is almost inevitable. Light is just asking to crash into something—and it does, all the time. A variety of things happen when light hits an object. The light can be reflected or scattered as it bounces off an object; it can pass straight through the object; it can be distorted, bent or refracted by the object or it can be absorbed by the object. Any combination of these things can happen, depending on the circumstances and the object in question. All of these things are good. If they didn't happen, we wouldn't be able to see.

Light travels in wave form with the different colors of light having different wavelengths. Light visible to the human eye comes in different colors (directly relating to the production of those different colored photons) from violet, indigo and blue to green, yellow, orange and red. When all of these wavelengths are combined, they produce white light. When you see light shining through a transparent object with two surfaces, such as a glass of water or a glass prism, the different wavelengths are refracted, bent or distorted to different degrees so that they separate and you can actually see the different colors of light. When light is refracted through water droplets in the air, the different colors can be seen in the form of a rainbow. The same effect gives a diamond a halo of color. In some cases, seeing the different colors will be the result of "interference" when distorted light waves and regular light clash to create different wave-lengths, allowing us to see different colors in the light.

The flash from the world's first atomic bomb test in the New Mexico desert was seen all over New Mexico, as well as in parts of Arizona, Texas and Mexico.

Light travels at a speed of 186,000 miles (299,350 km) per second—that's almost 670 million miles (1 078 million km) per hour!

To help you see clearly and discern different colors, each of your eyes has about 126 million sensors.

An explosion is basically a fire that burns very fast.

Normal glass allows light with wavelengths in the visible spectrum to pass through, as do a number of other mediums. Glass will also allow some light forms beyond the visible range (such as gamma rays or X-rays) to pass through, but lower-frequency light (such as ultra-violet or infra-red) will not penetrate glass. Because light is able to pass through glass, we can see outside, beyond the window. Light that has bounced off objects outside floods in through the window, bringing to our eyes the images of, for example, cars trundling along the road (rather slower than the speed of light) stuck in a traffic jam. So while those cars are standing still, let's take a look at them. The metal bodywork of the cars reflects light very well, the molecular structure of the metal and its smooth surface resist light penetration and reflect the light waves so that the cars appear shiny. An object with a smooth surface reflects light so well that, when viewed from an angle where the light is reflected directly towards your eyes, you can't really "see" the object at all. All you see is the bright, white-light reflection.

A rough surface, on the other hand, tends to scatter a certain amount of light. The paper that this book is printed on has

The way it reflects light is what makes an orange look orange

a "rough" surface, so you can look at it from almost any angle and still see the page and the words on the page. Another easy-to-see manifestation of the scattering of light is our sky. Because our atmosphere is "rough," filled with molecules of all sorts of different elements, some of the light waves from the blue end of the spectrum are reflected by particles in the atmosphere and scattered on their journey to earth. On a clear sunny day, therefore, the sky appears blue.

When light is absorbed by an object the light energy is not reflected but "soaked up" by the molecules of the object. Since energy cannot be destroyed, only changed into another form, some of this light energy will be transformed into heat. The way that an object absorbs light is just as important to how we see it as the way in which it reflects light. In fact, the only way we know that an orange is orange or a strawberry is red is by the way their surfaces either reflect or absorb different colors of light. An orange, you see, is not inherently orange. It appears orange to us only because it is absorbing some of the light wavelengths and reflecting those that present themselves to us as orange. For an object to appear pure white, it has to reflect all visible light and, at the other end of the scale, for something to appear black it has to absorb all of the light wavelengths.

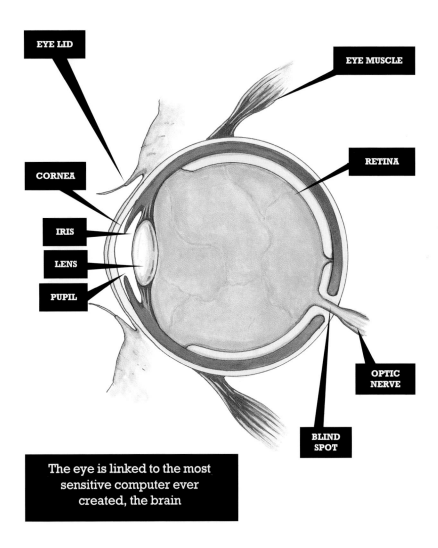

EYE LID

EYE MUSCLE

CORNEA

RETINA

IRIS

LENS

PUPIL

OPTIC NERVE

BLIND SPOT

The eye is linked to the most sensitive computer ever created, the brain

PARTY ANIMALS

RODS

HARD WORKERS

The 120 million rods on your retinas outnumber the cones by 20 to one

CONES

Our own sensitive sensors

With all of this physics and chemistry going on all around us, how on earth do we make sense of it all? There's stuff bouncing around at millions of miles an hour, invisible energy beams shooting all over the place—just how do we cope with it all? The answer is that we use the most advanced computer ever invented and stick part of it on the outside of our heads. The computer is your brain and the bits on the outside of your head—your eyes—are part of your brain.

Your eyes are immensely sensitive sensors, able to use their receptors to receive hugely complicated energy signals in the form of light, translate that energy into nerve impulses, like electrical signals, and transmit those signals through the optic nerve to the brain where they are decoded into images and colors.

To gain some idea of the complexity of this operation, take a look at your own eye. Once light has passed through the outer part of the eye, the lens focuses it onto the retina on the inside back wall. The retina is composed of millions of light-sensitive cells called rods and cones. There are 120 million rods in your retina, mainly around the edge of the retina and dealing primarily with black and white images. Cones are concentrated around the middle of the retina and process color images. There are about six million cones in your retina.

Rods are party animals that come into their own when the lights go down as they are better able to discern images in poor light, helping to explain why your color vision isn't so good when you bumble around in near darkness.

Cones, on the other hand, are hard workers, dealing with the heavy-duty labor of making sense of intense light energy that creates a confusion of color. Three different kinds of cones handle short, medium and long wavelengths so that the brain can discern the full range of colors.

Of course all of this happens a lot faster than you can even think about it. In fact, if you had to think about it, you wouldn't be able to process all of this information quickly enough to allow you even to walk across a room. You'd still be trying to figure out exactly how far away that cat (no, it's a dog; check again; affirmative, computer brain confirms this image relates directly to memory bank images labeled "dog") on the rug is from where you are standing and whether that dog on the rug is brown or green (green dog does not compute; check again; negative, computer brain identifies rug as green and dog as brown) when you trip over him. See how long it would take to think it through? And that's a lazy old green dog who hasn't moved off that rug all afternoon! With a bang and an explosive flash you really wouldn't have a clue what was going on.

Georgia Green, on the other hand, with only a basic perception of light and dark, was able to understand straight away that the flash she "saw" was something truly awesome. It would be quite some time before the secret of the nuclear test was revealed and she and Joe Wills learned just how awesome it really was.

THE GODS ARE ANGRY!

WHEN YOU HEAR THAT TELL-TALE RUMBLING bang that we know as thunder, perhaps you think: "Wow! That was loud!" or "That was so loud you could almost feel it!" or "I knew I shouldn't have eaten so much chilli last night!" You probably don't think: "The gods are angry with us! Run and hide!" But that is exactly what people used to think. The sound of thunder was the sound of the gods' displeasure. If you were from northern Europe, on the other hand, you cheered the arrival of Thor (also known as Donar, Thur, Thuner or Thunar), the Norse god whose great chariot made the noise of thunder as it trundled across the sky drawn by two goats. Thor, you see, brought the rain that was needed for the peasants' crops. Destruction from the thunderstorm would come only if Thor was having a bad day.

In the modern world, of course, we know that thunder is the result of lightning (see page 128).

Hot air

When a lightning bolt is triggered it causes the surrounding air instantly to become superheated and, like almost any other gas that is rapidly heated, the molecules of air expand, rushing outwards to fill a greater volume. In so doing, they collide with the more densely packed molecules of cooler air and the energy from the expanding "hot" air is passed on in the form of a shock wave. The shock wave then travels rapidly outwards from the vicinity of the lightning flash and it is this disturbance in the air that comes to us as the sound vibration of thunder. The fact that it is a shock wave, an energy wave, means that if you are close enough to the storm you will hear it as being very loud and you will be able to feel it, too.

So, it wasn't the chilli and, nor is there an angry Thor slinging lightning bolts, pounding the clouds with his great battle hammer and driving a goat chariot in the sky.

IF YOU WERE FROM NORTHERN EUROPE YOU CHEERED THE ARRIVAL OF **THOR** THE NORSE GOD WHOSE GREAT CHARIOT

MADE THE NOISE OF THUNDER AS IT TRUNDLED ACROSS THE SKY DRAWN BY TWO GOATS

FAMOUS BANGS

KRAKATOA

YOU HAD TO FEEL SORRY for distinguished actor Maximilian Schell when he appeared in the 1969 action-adventure movie *Krakatoa East of Java*. His ship was hijacked by villainous treasure hunters in the Indian Ocean while the entire island of Krakatoa erupted and a giant tsunami tidal wave wiped out everything in its path. He could be forgiven for not appearing to know whether he was coming or going, however, Krakatoa was, and still is, *west* of Java, in the Sunda Strait between Java and Sumatra. Just to add to poor Maximilian's confusion, the movie was actually filmed in Majorca in Spain!

Despite the geographic anomalies, the movie was based on historical fact. Krakatoa, a tiny volcanic island, erupted August 27 1883 with what was one of the biggest bangs ever heard by humans. The explosion was audible over a distance of about 3,000 miles (4,800 km).

Explosive force

The volcano erupted with the explosive equivalent of 100 megatons of TNT, hurling about 5 cubic miles (over 20 cubic km) of material up to 30 miles (48 km) into the air. At the higher altitudes, the ash and dust circled the equator in just 13 days. Jakarta, over 100 miles (160 km) from the site, was plunged into darkness as the spreading cloud of dust blocked out the sun completely. Ultimately, ash and dust would settle over an area of more than 300,000 square miles (777,000 km^2), but airborne particles caused three years of remarkable red sunsets as far away as New York. So much dust was hurled into the atmosphere that it diluted the power of the sun, lowering average temperatures by 1.8°F (1°C). Temperatures would not return to normal for nearly five years.

The eruption generated a huge wave, or tsunami, that reached a height of about 130 feet (40 m) as it rushed ashore, devastating more than 150 coastal villages on surrounding islands and killing over 36,000 people in the process. Huge blocks of coral weighing as much as 600 tons (610 tonnes) were deposited on the shoreline by the wave which, although much reduced in size and power, reached Aden (4,000 miles/ 6,400 km away) in just 12 hours—a sea journey that would take most ships well over a week.

At the center of all this destructive power, the little island of Krakatoa tore itself apart. Having started off just over 3 miles wide by 5½ miles long (5 x 9 km), the island was reduced to one-third of its previous size, although the volcanic activity did produce some new islands nearby where the sea had once been 100 feet (30 m) deep.

The Krakatoa eruption was one of the world's biggest ever and—thanks to Hollywood's portrayal of the event—most famous bangs.

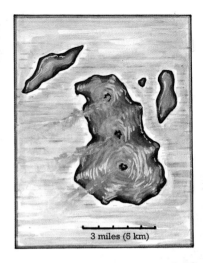

Prior to the volcanic eruption in 1883, the island of Krakatoa was 3 miles (5 km) wide by 5½ miles (9 km) long

3 miles (5 km)

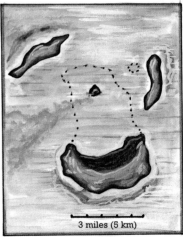

The eruption blew the island to pieces, leaving it one-third of its original size

3 miles (5 km)

Although modern car engines are more sophisticated, they still rely on the same basic principles as the earliest internal combustion engines developed over a hundred years ago

BANG GOES THE ENGINE

A BANG—A FLASH OF FLAME, a puff of smoke and a loud noise—in a quiet residential street. What was it? A gas mains explosion? A gun shot? A bomb? And why is no one paying the slightest bit of attention? It was an explosion wasn't it? It was, but no one pays any attention because they are so used to these explosions. These bangs happen up to a hundred times or more every second as a normal car drives along the street. They happen inside the car's engine. They are what provide the power to turn the car's wheels.

The internal combustion engine

Most cars today, despite great advances in alternative technology, are still powered by engines using the same basic principle that powered the first internal combustion-engined cars over a hundred years ago. They rely on a controlled explosion to provide the power. The explosion happens inside a cylinder, of which the car's engine may have many, although most commonly there will be four or eight.

To understand how the cylinder controls an explosion to produce motion that turns the wheels of the car, first think of how an artillery gun works. There is a cylinder, which is the barrel of the gun, and there is an explosion that sends the artillery round hurtling up the cylinder and out of the open end in the direction of the target. Now imagine that the open end of the gun barrel is sealed shut. Not only is it sealed shut, but the seal is actually a powerful spring-loaded platform. Now, when the artillery round reaches the end of the barrel, it doesn't head off in the direction of the target. Instead, it hits the spring device and goes back down the barrel. Then, just as it gets back to its starting point another explosion goes off and the whole process starts again, with the artillery round shuttling back and forth up and down the barrel.

In a car engine, the cylinder contains a piston that performs much the same sort of shuttling act as our imaginary artillery round. Almost all car engines work on what is known as a "four-stroke combustion cycle," the four different stages of the combustion cycle being intake, compression, combustion and exhaust.

1 INTAKE
Here the piston starts at the top of the cylinder. The inlet valve opens and a mixture of fuel and air is drawn in as the piston moves down the cylinder. This is a misty spray of, in the case of a petrol engine, gasoline so fine that when mixed with air it is really a gas rather than the liquid with which you filled your vehicle's tank.

2 COMPRESSION
With the inlet valve closed, the piston now moves back up the cylinder, squeezing the gasoline and air gas into a small space under pressure, thus ensuring that maximum power will be achieved when the mixture explodes and expands.

3 COMBUSTION
With the mixture in the cylinder now under maximum pressure, a spark is created by a spark plug. This ignites the mixture and the piston is driven down the cylinder again by the explosion.

4 EXHAUST
When the piston is at the bottom of the cylinder, the exhaust valve opens and waste gases are drawn out as the piston returns to the top. The exhaust valve then closes, the intake valve opens and the whole process starts again.

Synchronized cylinders

Of course, the piston does not shuttle back and forth freely inside the cylinder. It is attached via a connecting rod to a crankshaft which, via a series of gears, turns the vehicle's wheels.

The accelerator in your vehicle determines how much fuel is fed into the cylinder and, therefore, how powerful the explosion, how fast the piston travels down the cylinder and how fast the crankshaft is then turned.

In an engine with four cylinders, the different stages of the combustion cycle are synchronized so that the crankshaft is constantly being driven round—one of the cylinders performs its combustion stage while the others are either in exhaust, intake or compression mode. In an eight-cylinder engine the cylinders will be synchronized so that two perform each stage together, thus easing the amount of work that each cylinder has to do and the amount of power required from each combustion or explosion. Conversely of course, when the eight are all working hard, even more power is produced.

So, if you see an old lady driving to church on a quiet Sunday morning, just think how many millions of explosions she has set off by the time she reaches the parking lot. It certainly makes Sunday mornings go with a bang!

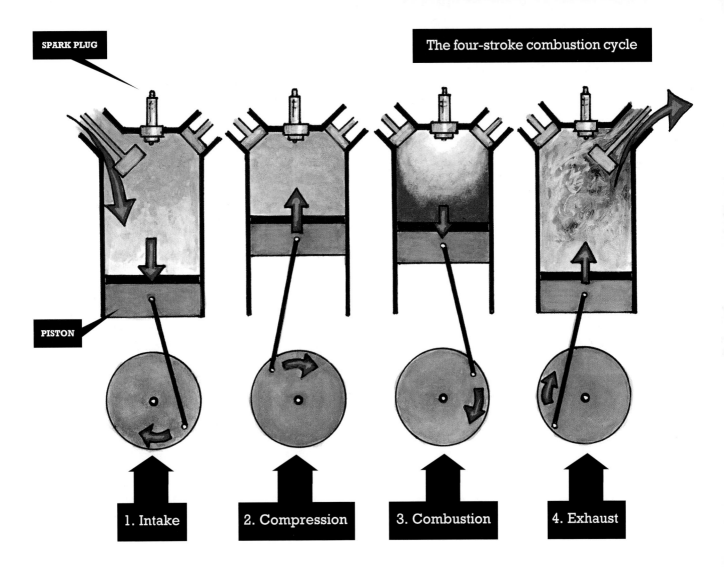

SPARK PLUG

PISTON

The four-stroke combustion cycle

1. Intake

2. Compression

3. Combustion

4. Exhaust

BLOWING YOU AWAY

THE DESTRUCTIVE FORCE OF A BIG-LEAGUE BANG, a major explosion, is enough to take your breath away... and most of your other bits and pieces as well. But just exactly how does a big bang cause so much devastation—or indeed any devastation at all? We talk about things being "blown up," "blown to pieces" or "blown away," but why "blown"? That's what the wind does. What's blowing got to do with an explosion? Quite a lot actually. A big bang is the worst case of wind you'll ever come across.

Faster reactions mean bigger bangs

As we learned in "Bang To Light" (see page 24), where we looked at how an explosion creates a flash of light, any explosion is really just a kind of fire. It is not, however, the kind of cosy fire you'd want burning in your fireplace while you toast marshmallows. The key thing to remember is that this is a fire that can burn itself out faster than the quickest flicker of flame from that log in your fireplace. It is a far more violent reaction than the gentle chemical transformation going on as your logs slowly burn in the grate, but essentially the same thing is happening. The burning logs are actually changing form and turning to gas and other waste products, with the different gases produced being warmed by the heat and drifting lazily off up the chimney with the rest of the components of the smoke. The combustible matter in explosives also changes form and turns to gas. The power of the explosion, the size of the bang, all depends on how quickly the chemical reaction (in a conventional explosive) takes place. The faster the reaction, the more violent the result. Needless to say, the amount of explosive used affects the size of the bang, too but, pound for pound, an explosive compound that reacts faster will produce the bigger bang.

Detonation

Once an explosive charge is detonated, either by applying heat, an electrical charge or some other form of energy (just like starting a fire with a match), the chemical reaction begins. The gas produced wants to expand and occupy a far bigger space than the original solid explosive. It pushes everything out of its path and forms a shock wave within the explosive material—a surge of energy with the gas molecules energizing all the other molecules of the original explosive matter. This causes them to change and expand in a kind of chain reaction that doesn't stop until all of the combustible material—the explosive—has been used up. In other words, it is like fire spreading through a log or pile of logs, burning up everything until there is nothing left to burn. The big difference is of course that with explosives the "burning" is almost instantaneous.

The instantaneous nature of the reaction means that the gas into which the explosive transforms doesn't drift lazily off up a chimney into the atmosphere; it doesn't have time. It's like a bunch of hyperactive kids spilling out of school at home time. With all of the new gas molecules trying to occupy the same

A big bang will produce a wall of high-pressure air like a gale force wind blowing in all directions

The new gas molecules produced by the detonation push and shove and try to occupy the same space—just like kids spilling out of the school yard at home time—eventually bursting out in a shock wave of energy

space in the school yard at the same time, there's a lot of pushing and shoving going on and even more of the little bullies arriving all the time. Eventually, they burst out of the yard altogether, pushing everything out of the way. A shock wave of energy surges through the explosive and out into the immediate surroundings. All of this happens in the tiniest fraction of a second.

Depending on the type of explosive, the shock wave might consume the explosive material at a speed of about 18,000 miles (29,000 km) per hour. This is roughly the rate of expansion of gases created when the explosive C4 detonates, and equates to 5 miles (8 km) per second. This wave of energy erupts in all directions. Just like the ripples created when you drop a pebble into a pond, the energy wave rushes outward, although it will go up and down as well as sideways, an expanding ball of energy compressing the air in front of it and driving a wall of high-pressure air in all directions.

The power of the shock wave

Anything in the immediate vicinity of the detonation will either become part of the chemical reaction of the explosion and contribute to the effect or be pushed outward by the shock wave. Any stones or loose material near the explosion will, therefore, fly through the air as a deadly and destructive missile. This is why terrorist bombs sometimes have ball bearings or nails packed around them. These will be hurled outwards at thousands of miles an hour as lethal shrapnel.

Larger objects will be hurled away from the explosion, too—bricks and furniture, even cars or trucks depending on the quantity of explosive that has been detonated. Such flying debris can and does cause massive damage to structures such as buildings but so can the shock wave itself. When the shock wave hits something that refuses to move, like the wall of a building for example, all of the energy has nowhere to go and the pressure that immediately builds up can simply push the wall over. It is like an invisible elephant leaning on a yard fence—or rather falling against it, as the shock wave travels so fast that it generally arrives as a punch of energy rather than a sustained push. Eventually, the energy will expend itself and the effect of the shock wave will peter out like a dying breath of wind, but by then, depending on the geography of the area, it can have covered quite some distance. After the Trinity atomic bomb tests in New Mexico in 1945, it was reported that the shock wave caused windows 120 miles (193 km) away to break.

The shock wave doesn't just travel through the air of course, it will radiate outwards through the ground, too. Although this form of shock wave will travel more slowly and is unlikely to be quite as destructive as the airborne version, it can still contribute to the overall devastation. Acting and feeling just like an earthquake, the shock wave traveling through the ground will cause buildings to rumble and shake. If there was anything left of that fence after the invisible elephant finished with it, then the slower-moving ground shock wave will arrive an instant later to shake it to pieces. Indeed, some bombs have been developed specifically to take advantage of the earthquake effect.

Man-made shock wave devices

During World War II, British engineer Barnes Wallis developed a number of bombs for the Royal Air Force including the Tallboy and Grand Slam devices. Both of these bombs were designed to penetrate concrete, their primary targets being heavily fortified German submarine pens or V-1 rocket launch sites. Dropped from a Lancaster bomber flying at 20,000 feet (6,100 m), the Tallboy could penetrate up to 16 feet (5 m) of concrete. More than half of its 12,000 lb (5,450 kg) weight was in the bomb itself, although it did pack a mighty punch from its 5,200 lb (2,360 kg) of explosive. It would hit the ground traveling faster than the speed of sound, plow into its concrete target and then detonate, sending its shock wave through the concrete structure in an attempt to shake it to pieces. If it passed straight through the roof and detonated inside the structure, of course the effect would be just as devastating. Tallboy was actually known as "the earthquake bomb." Towards the end of the war, Wallis created Tallboy's big brother, Grand Slam. At 22,000 lb (10,000 kg), Grand Slam was almost twice the weight of Tallboy and when it hit the ground it created a crater 100 feet (30 m) deep and more than 100 feet (30 m) across. It remains the biggest conventional bomb ever used in anger.

The effect of the shock wave to "blow things away" was ably demonstrated by another bomb, this time developed for the U.S. military. Known

Explosions are basically the same kind of reactions that go on in your camp fire

as the "Daisy Cutter" when used in Afghanistan, and as "Commando Vault" when originally deployed in Vietnam, this is one of the largest conventional bombs in existence. It is designed to detonate its 12,600 lb (5,720 kg) of explosive only a couple of feet (0.6 m) above ground level. The shock wave created flattens everything within a radius of 100—300 yards (91—274 m). It was used in Vietnam to clear helicopter landing zones in the jungle.

The hugely destructive effect of the shock wave has also been demonstrated far from what we would normally term "battlefields." On March 12 1993 a van loaded with home-made explosives was blown up in the underground parking lot of New York City's World Trade Center. Six people died in this first attack on the tragic Twin Towers. The explosion caused an intense fire and $1 billion worth of damage to the building. Another truck bomb was set off a few weeks later in Bishopsgate in the heart of London's financial district, where one person died and over 40 were injured. The shock wave caused structural damage and shattered glass in dozens of buildings with a medieval church, St Ethelburga's, completely collapsing.

Vacuum effect

With the power of the shock wave deriving from the speed of its creation, and the gases created in the explosive reaction evacuating the immediate vicinity of the explosion at such a fantastic velocity, another interesting effect is created. Evacuation of the site of the explosion actually creates a vacuum. The vacuum then sucks all sorts of material back into the point of detonation. The invisible elephant may have flattened the fence, and the ground shock wave shaken it to

The faster the chemicals in any explosive react with one another, the more violent the result.

Gases expand from a detonation of C4 explosives at a rate of 5 miles (8 km) per second.

After the first atomic bomb tests in New Mexico in 1945, the shock wave was said to have broken windows 120 miles (193 km) away.

The 22,000 lb (10,000 kg) Grand Slam bomb of World War II created craters 100 feet (30 m) deep and more than 100 feet (30 m) across.

BARNES WALLIS

1887–1979

A DOCTOR'S SON, Barnes Neville Wallis was born in Derbyshire, England, on September 26 1887 and moved to London with his family when he was two years old. On leaving school, Wallis worked for Thames Engineering where they made engines for ships but by 1913 the young engineer was working for Vickers on the construction and design of airships.

At the outbreak of World War 1 in 1914, Wallis volunteered for the army, sneaking a close-up look at the sight test chart prior to his test and memorizing it to ensure that he wouldn't fail. The authorities decided, however, that his skills were required designing airships and he worked on in the aircraft industry.

Britain needed his engineering skills once again, 25 years later, when World War II began. He designed the Wellington bomber and the Tallboy "bunker buster" bomb, but is most famous for his invention of the "bouncing bomb" used to destroy German dams and help cripple their industrial production.

Wallis continued working in the aerospace industry after World War II, contributing a constant stream of ideas for things such as swing-wing and supersonic aircraft even beyond his retirement at the age of 83. He was knighted in 1968 and died at the age of 92 in 1979.

pieces but the explosion also created a vacuum cleaner to tidy away all the broken bits! If that sounds ever so slightly ridiculous, then consider the story told by one London fireman during World War II when the city came under attack from the sinister V-2 ballistic missiles. The V-2 arrived at its target faster than the speed of sound, so no one heard it coming—the first you knew of a V-2 attack was when the explosion happened. This meant that it was more than a bit confusing for the London fireman who was on duty inside the station when a V-2 struck nearby. The shock wave blew open the massive front doors of the station before he knew what was going on. Then, just as quickly, the backdraft from the vacuum slammed them shut again!

Medicinal shock waves

The destructive power of shock waves has recently been harnessed to produce results far more beneficial to mankind, especially to those afflicted with painful kidney stones. Acoustic shock waves can be created electrically and directed at the kidney stones to break them up. It doesn't blow you away, just little bits of you that you don't really want!

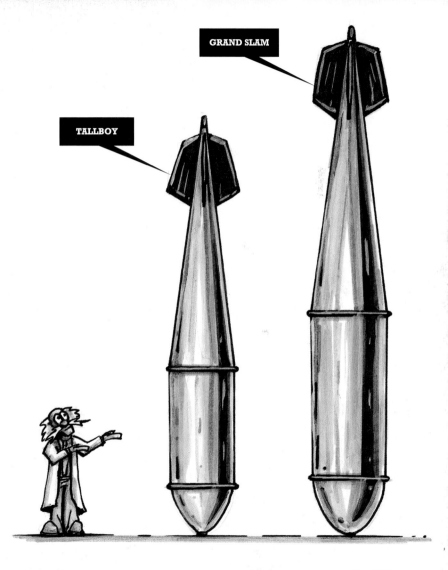

GRAND SLAM

TALLBOY

THE DEADLY CUSTARD BOMB

It's in almost everyone's kitchen cupboard, it forms a major part of the diet of almost all of the world's population, it must surely be regarded as one of the most innocuous substances known to man, and yet, it is a deadly explosive! What is it? It's custard powder.

OK, so not everyone keeps custard powder in their cupboard and custard is certainly not eaten on a regular basis by almost everyone in the world. The major ingredients of custard powder, however, are cornstarch and flour; in fact grain in general fits all of the above criteria—including the bit about being a deadly explosive.

If you're not convinced of this, try the experiment overleaf. Although it is not an entirely reliable experiment, if you practice and fine tune it, you will probably be able to make it work. You may well also lose your eyebrows. And for goodness sake never suck on the straw! Why, though, should custard go bang at all?

Why does custard explode?

White flour consists largely of starch and starch is a carbohydrate. On a molecular level this means it is made up of chains of sugar molecules. Sugar, of course, burns rather well and since any explosion is basically a quick-burning fire, the explosive potential now becomes a little clearer. It's not simply a case of setting light to a mound of custard powder, though. If you were to take a pile of

custard powder or flour and stick a match in it, all that would happen is that the match would go out. Where fine particles of flour or grain dust are wafting about in the air, however, the situation changes entirely.

Many carbohydrates become highly flammable when, as dust, they are exposed to a naked flame. In fact, a particle of dry flour dust will flare and burn incredibly quickly. In the tiniest fraction of a second it will ignite and set fire to the particle nearest to it, which will in turn set fire to the particle nearest to that, causing an incendiary chain reaction. The ever-expanding ball of flame

flashes through the dust cloud, causing— especially in an enclosed space where pressure builds up from the expanding gases produced by the burn—a massive explosion. That's what blows the lid off the coffee tin in our experiment overleaf, but when the same thing happens on a larger scale the results can be catastrophic.

Grain dust explosions

In Wichita, Kansas, on June 8 1998, seven people were killed and 10 injured when a grain elevator exploded. The force from the blast could be felt over a wide surrounding area and much of the installation was destroyed. Army helicopters and a 300-foot (90-m) high crane were used to rescue some of the workers from the roof of the complex amid continuing fears that rescue attempts could actually instigate another explosion. Grain dust explosions are far from uncommon, with an average of over a dozen every year over the past few years in America alone. The major problem in a grain processing facility is that dust forming in the air can be ignited by a heat source or a spark from machinery or even an electric light. The initial dust blast then stirs up an even greater cloud of grain dust, which explodes an instant later, ignited by the flash from the first explosion.

The deadly custard bomb may provide a surprise bang for fun, but when one of its bigger brothers goes off it is no laughing matter.

EXPLODING CUSTARD

You will need:

A LARGE COFFEE TIN
A LONG STRAW OR SOME THIN TUBING
A SMALL CANDLE
SOME CUSTARD POWDER

1 Punch a hole in the side of the coffee tin near the base. Feed your straw or thin tubing in through the hole. It should be a tight fit.

2 Stand the coffee tin on a firm base—a table or work bench—and place your candle inside the coffee tin.

3 Spoon a small mound of custard powder over the end of the straw or tubing inside the tin.

4 Light your candle then quickly fit the lid of the coffee tin on tightly. Quickly blow a puff of air through the straw. There will be a loud bang and a flash as the lid is blown off the coffee tin.

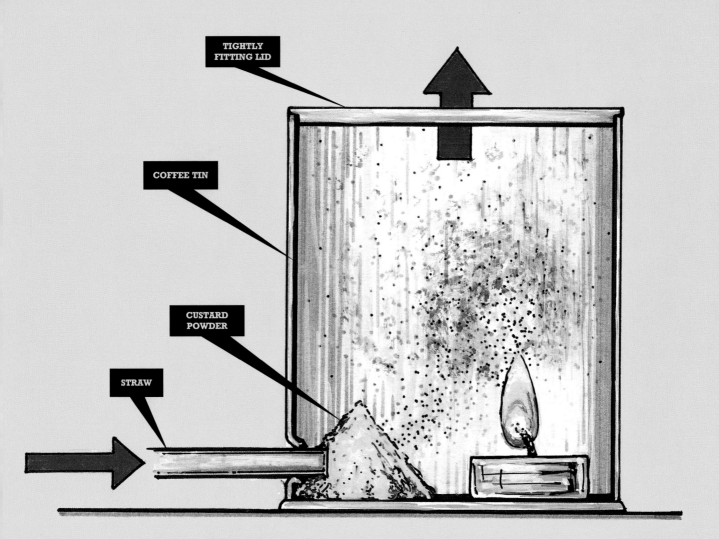

TIGHTLY FITTING LID

COFFEE TIN

CUSTARD POWDER

STRAW

WHY BALLOONS GO BANG!

THE BURSTING OF A BALLOON is probably one of the first really loud noises that most of us experience at pretty close range. Little kids love to play with balloons, batting them around, bashing them with toys, and biting them... There is nothing as priceless as the expression on a small child's face just after he or she has fallen on or experimentally bitten into an ordinary party balloon and it has gone off with a bang. The loud noise and the rush of air from the bursting balloon take little kids, and most of the rest of us, completely by surprise. There is really no other way for children to produce quite such a startling noise until they learn to blow up paper bags and make them go bang. The same principle, of course, is at work to produce both balloon and paper bag bangs. But why does a balloon go "bang" and not "wheeeee," "tinkle" or "snigger"? The answer, my friend, is blowing in the wind.

Under pressure

Pressure is the key to the balloon-burst bang. Inflating a balloon requires air pressure, and anyone who has almost popped their eyeballs trying to blow up those long skinny party balloons knows just how much pressure is required. The rubber, or latex, of the balloon stretches as air is forced in. The balloon "contains" the air under pressure, forcing it to remain compressed when what it really wants to do is spread out among the surrounding

PUFF PUFF

air at normal atmospheric pressure. It will do this if there is a hole in the balloon. A pin hole put in a balloon before it is inflated will probably result in the balloon slowly deflating once air has been blown in. The air will escape to the atmosphere quite slowly through the pin hole which, as the balloon was inflated, will have retained its shape, the rubber around the edge of the hole staying quite strong. The pin hole will remain recognizably a pin hole once the balloon has entirely deflated.

If you suddenly introduce that pin hole to an inflated balloon, however, the result is markedly different. The pressurized air inside the balloon rushes to escape, rapidly expanding the pin hole, suddenly meeting the lower-pressure air on the outside and creating a high-pressure shock wave that rips the balloon apart. Take a look at a deflated balloon that went "bang" and you

won't find a neat little pin hole. In fact, there may be bits of that balloon you can't find at all!

Why "bang"?

A major part of the "bang" sound involves the way that we hear noises. We hear sound as vibration, including those sounds that we make ourselves by vibrating our vocal chords. The sound energy is transmitted by vibration through all sorts of media—solids like wood, metal or concrete; liquids like water, and even air. The vibration is passed from one particle of air to another in a wave motion that eventually impacts on our ear-drums causing them to vibrate. This membrane then transfers the vibrations to the inner ear where the energy of the vibrations are translated into nerve impulses and trans-mitted onwards to the brain for interpretation.

The shock wave set up by the high-pressure air rushing out of a popped balloon creates a high-energy burst that hits our ear-drums like a drumstick hitting a drum. This explains why we hear the bursting of a balloon as a bang—and we very quickly learn not to bite them!

The energy used to blow up these balloons is stored in the high-pressure gas, ready to be converted

EL ALAMEIN

WITHOUT DOUBT, one of the loudest, most sustained, most deliberate and most terrifying man-made bangs ever heard happened in the desert of North Africa on the night of October 23 1942. This was the artillery fanfare that announced the commencement of one of the key battles of World War II—El Alamein.

Round one

In fact, this was the second battle of El Alamein. The first had been fought in July with "the Desert Fox," German General Erwin Rommel, who pursued an Allied army of British, Australian, New Zealand, Indian and other troops across hundreds of miles of unforgiving terrain before the final showdown in the western desert. The Allied commander, General Claude Auchinleck, prepared his defenses along a 30-mile (48-km) front, which stretched from the Qattara Depression in the south (an area of salt marshes almost impassable for tanks and other heavy armor), through a central area of barren rocky ridges and escarpments towards the

Mediterranean, via a sandy coastal plain in the vicinity of a small railway halt called El Alamein.

Both the German Afrikakorps, along with their Italian comrades, and the Allied armies were at a low ebb after many months of fighting. Auchinleck could call on 35,000 men and only 160 tanks to make his stand against the might of the Afrikakorps, which had swept all before it during the North African campaign. Little did Auchinleck know that Rommel was in desperate need of reinforcement. Over 1,000 miles (1,600 km) from his main base, Rommel's supply lines were stretched to the limit, resupply was further hampered by Allied attacks on naval convoys crossing the Mediterranean and at the spearhead of his advance he had only 60 German tanks, 30 Italian tanks, and less than 7,000 troops.

Nevertheless, Rommel decided that a bold attack was called for. The attack failed, and Rommel was forced to establish a battle line facing Auchinleck's El Alamein defenses. Offensive and counter-offensive followed over the course of the next month until, with all of the combatants exhausted, the first Battle of El Alamein ground to a halt. By then Rommel's force was reduced to just 26 tanks and Auchinleck had sustained 13,000 casualties.

Round two

The arrival of a new British commander, General Bernard Montgomery, along with massive Allied reinforcements and the departure of Rommel for medical treatment in Germany set the scene for the second battle. To give some idea of the scale of this conflict, "Monty," as the new Allied commander was known, had at his disposal a massive arsenal of 1,350 tanks and 500 aircraft along with 220,000 troops. Facing them, also reinforced and having prepared a 45-mile (72-km) wide defensive line behind a double row of minefields, was Rommel's army, now consisting of about 500 tanks and 110,000 men.

The numerical advantage was with the Allies and Monty knew that he would need all of this strength to win through. Accordingly, he began the assault with the biggest artillery strike launched against an enemy since World War I. It has been described as a 1,000-gun barrage (although there were actually slightly less than 900 guns) that made the earth tremble beneath the feet of the Allied gunners. Although the offensive faltered along the way, within two weeks the Allies were able to claim a resounding victory. The cost in lives was immense, with over 70,000 casualties in total, but this battle did stand as a turning point in World War II, bringing the Allies a much-needed victory after three long years of war.

The beginning of the end

The end of the battle was swiftly followed by the landing of thousands more Allied troops, including U.S. servicemen, which brought the North African campaign to an end. The Allies' strong position in North Africa then served as a springboard for the invasion of Sicily and Italy on the other side of the Mediterranean Sea and the beginning of the end of World War II. Although there would be three more years of fighting, the big bang at El Alamein remains one of the most decisive actions of the entire war.

DANGER— EXPLOSIVES!

YOU MAY WELL HAVE SEEN SIGNS warning of danger from explosives if you live near some kind of mining facility or a military base. Trucks carrying such materials are required to show a hazard warning notice to let everyone know that they have a dangerous cargo. This is of special interest to the emergency services in the event of a road accident. They don't want to make an accident site a thousand times worse by mistreating a dangerous load that may already have become unstable during the accident. This sort of situation is of course far from commonplace, but what about the explosives themselves? What is an explosive and how commonplace are they? Surprisingly, explosives are all around us!

What are explosives?

To define an explosive, you first have to think back to how we defined an explosion in "Bang To Light" (see page 24). Were you paying attention? No? OK, basically an explosion is a fire that happens really fast. Almost anything that burns is capable of producing some kind of explosion under the right circumstances. Even wood, which burns slowly enough for us to be able to sit safely right by a log fire and warm our hands, can produce an explosion. Sawdust particles hanging in the air can ignite, causing an explosive reaction just like the grain dust described in "The Deadly Custard Bomb" (see page 48). You weren't paying attention then either? Skip back a few pages and have a read. We'll wait for you…

Ready? Good. There are of course more obvious explosive materials around us than simple sawdust. Gasoline (petrol), all kinds of fuel oil and the natural gas used in our cookers and heating systems spring immediately to mind but even they are not what we would view as purpose-built explosives, things that are

Rail networks worldwide were built using explosives to blast routes through difficult terrain

designed to go off with a bang and blow stuff apart. The most basic, and the very first, man-made explosive was black powder—known as gunpowder.

Black powder—the first explosive

There are various theories as to how gunpowder was invented and by whom, but most people tend to believe that it was first developed in China in the ninth century. Legend has it that black powder was invented by a Chinese cook, who mixed the three main ingredients that he may well have had lying around in his kitchen. Black powder is 75 percent potassium nitrate (saltpeter—a salt substitute), 15 percent charcoal and 10 percent sulfur. That doesn't make it sound like he was concocting something that would taste very nice but, on the other hand, he may have been experimenting to try to find some form of fire lighter.

The black powder is what is termed a low explosive. In the open it will burn extremely vigorously, the sulfur igniting easily, the charcoal burning to feed the flame and the potassium nitrate providing a self-contained oxygen supply. The Chinese cook, therefore, might well have found that this could help him start his cooking fire. In an enclosed space, of course, the low explosive will go off with a bang—a very big bang if enough of it is used. That poor old Chinese cook may well have discovered this when a spark got into his pot of fire lighter powder. His recipe survived, though, even if he didn't, and within 300 years or so the Chinese were packing black powder into bamboo tubes and igniting it to hurl rocks as a primitive type of cannon.

BANG

Uses for black powder

Marco Polo traveled to the East and brought the black powder back to Europe in the 13th century. Initially it was used for decorative firework displays, but before long the military had found a use for it. By the 14th century cannon were being deployed on the battlefields of Europe.

The non-military uses of black powder took much longer to develop—about 300 years longer, in fact. It may have been used in mining operations in Germany and Czechoslovakia in the early 17th century, and by 1700 the art of controlling the powder to use it as a blasting tool in mines was well advanced. One of the first examples of its use in civil engineering was in the excavation of the French Canal du Midi's Malpas Tunnel in 1679. Black powder began to be replaced by high explosives towards the end of the 19th century but is still used in some forms today in quarrying and blasting.

The most widespread use of black powder was, of course, as gunpowder in ammunition for all sorts of artillery and firearms. Its low explosive qualities—a relatively slow burn producing a gradual expansion of gases and consequent gradual increase in propellant power—made it ideal for blowing bullets out of the barrels of guns. A high-explosive charge trying to do the same job would explode too fast and would run the risk of blowing the gun apart rather than just firing the bullet.

The first high explosives

So what is a high explosive? As its name suggests, a high explosive converts its combustible content to gas at a high rate. The fast burn means that the gas is produced and expands at a much faster rate than it does with a low explosive like black powder. This high rate of expansion means much higher pressure is created, giving much more power from the blast. Needless to say, this can also make high explosives much harder to control. The first high explosive was developed by Italian chemist Ascanio Sobrero in 1846. Sobrero added glycerol to concentrated nitric and sulfuric acid to create nitroglycerin.

Nitroglycerin was (and still is) an extremely powerful explosive, but it could also be quite unstable. Pure nitroglycerin is an oily, colorless liquid, which is extremely sensitive to shock waves (don't drop it), heat (no smoking, please!) and just about any other form of energy input. Its reaction will generally be to produce an extremely aggressive bang. To add to Sobrero's problems in dealing with his new

Was gunpowder invented by a Chinese cook trying to create some kind of fire lighter?

discovery, it had a high freezing temperature of about 55°F (13°C). This meant that his liquid would turn into a solid on a cool day—and the solid was even more volatile than the liquid. When "nitro" goes bang it generates gases that will instantly expand to over a thousand times the original volume of the nitro, causing an energy shock wave that travels at 8,420 yards (7,700 m) per second—about 17,000 miles (27,400 km) per hour.

Sobrero's creation needed further development to create a more stable, more user-friendly explosive and by the 1860s Swedish scientist Alfred Nobel had come up with one answer. He decided that mixing liquid nitro with something relatively harmless like charcoal or sawdust that would soak it up, would make the explosive easier to handle. In perfecting his concoction, Nobel invented dynamite.

Modern explosives

Other forms of high explosive have come along in the ensuing 150 years or so—two of the best known being TNT and C4.

TNT (trinitrotoluene) is a yellowy material produced by combining toluene with sulfuric and nitric acid. Toluene is extremely useful stuff. As well as being

Explosives are widely used today in mining, quarrying and civil engineering projects

highly flammable, it is used as a type of solvent, in adhesives, in pharmaceutical products, as a disinfectant, and in aircraft fuel (it is actually derived from petroleum). Because so much research has been done into the properties of TNT and its explosive effects, it has become the benchmark for judging the potency of most other explosive materials. In describing an explosion you may well hear someone say that it was the equivalent of 10 lb (4.5 kg) of TNT, or 500 lb (227 kg) of TNT or, in the case of a nuclear explosion or volcanic eruption, so many megatons of TNT.

C4 is so-called because it is a lot easier to say than cyclotrimethylenetrinitramine, which is the name of the explosive used in C4. C4 is a plastic explosive developed for the military. You may well hear C4 being mentioned in news reports on TV. C4 was used in the terrorist attack on the *U.S.S. Cole* in 2000, which killed 17 sailors. C4 was also the explosive one man smuggled onto an airliner in 2001, hidden in his shoes. Fortunately he had no way of effectively detonating the C4 because this, like other "plastic" explosives is quite a stable material. The make-up of C4 is such that it can be molded. It has the consistency of plasticine and the elements added to the cyclotrimethylenetrinitramine that give it its malleability coat the explosive to make it less sensitive to heat and shock. C4 needs a blasting cap or detonator to set it off— a small, controlled explosion that will then trigger the main explosive reaction. In fact, such is the power required to detonate C4 that it can actually burn without exploding. Soldiers in Vietnam used to set light to small pieces of it to make camp fires. When C4 goes, though, it goes with a real bang. Its rate of expansion is over 8,750 yards (8,000 m) per second—1,000 miles (1,600 km) per hour faster than nitro.

The beauty of being able to mold C4 is that a "shaped charge" can be created, directing the force of the explosion to where it is actually required rather than letting so much of the energy blast away in all directions. It can also be jammed into cracks in rocks or wound around steel beams to produce the desired effect.

The desired effect of course does not have to be in a military situation. High explosives are used today in mining, in demolition and in clearing obstacles for civil engineering projects such as roads, railways, tunnels, and bridges.

Perhaps one of the most bizarre uses, though, is for nitroglycerin. It is given to people with serious heart conditions. No, not in the sense of: "Here. Bite really hard on this and all your troubles will be over," but to ease cardiac pain by working as a medicine—a vasodilator. It expands blood vessels to make the heart's job of pumping blood through them a little easier.

Perhaps that Chinese cook in the ninth century was looking for a cure for heart disease rather than a handy fire lighter!

BANGS IN SPACE

WHEN THE SPACE SHIP of the evil green-skinned alien with three nostrils (you wouldn't want to meet him when he has a cold!) comes under attack from the fresh-faced space hero in any sci-fi movie or TV adventure, their laser weapons cause spectacular explosions in space as fast as the special effects people can produce them. As the battle rages there are bangs galore, providing enormous fireballs and thunderous bangs in glorious surround-sound stereo to crank up the excitement. The roar of a spaceship's engines as it thunders past is surpassed only by the sound of it being blasted to pieces, but how true to life are these space battles? What actually happens when a big bang goes off in outer space?

Soundless space

In the case of our green-skinned friend and his chums, the pyrotechnics of the space battles that appear on screen might not work in quite the same way in real life. For a start, if we were watching from our own little spacecraft (you be the pilot, I'll man the laser cannon), there would be no roar of a spacecraft's engine, no matter how gigantic that engine happened to be. Sound cannot travel through a vacuum and space is a vacuum—well, almost! There are a few particles of matter in every cubic inch of space, but it is nowhere near as densely packed as, for example, the air molecules and other elements in Earth's atmosphere. Without these particles to pass on the vibration of a sound wave, sound simply cannot travel. A spaceship would, therefore, pass right by us without us hearing the slightest growl from its mighty interstellar engines. Similarly, we wouldn't hear much of a bang if it were to explode. In fact, we wouldn't hear any bang at all, but there would be plenty of other evidence to show that it had blown up.

With no gravity or air resistance to shape it, an explosion in space would take the form of a sphere.

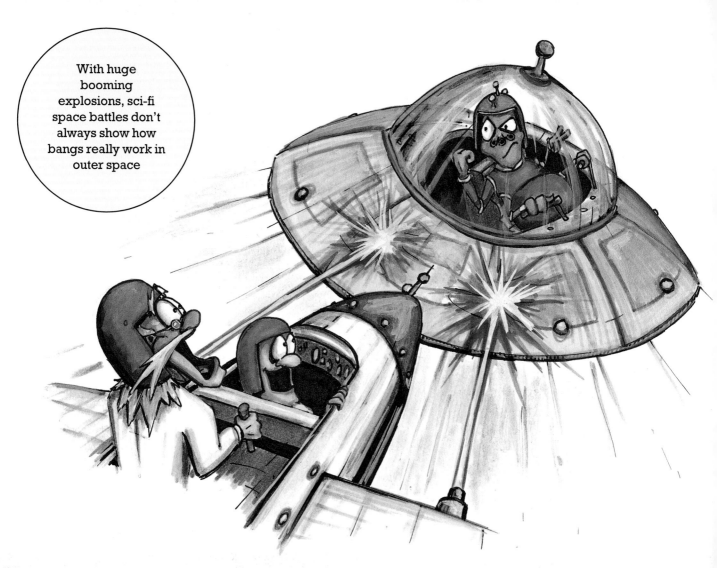

With huge booming explosions, sci-fi space battles don't always show how bangs really work in outer space

Great balls of fire

In the zero-gravity vacuum of space, the explosion would take on a quite bizarre form. Some might argue that, as there is no air and no oxygen in outer space, there would be no fireball, no great eruption of flame from the doomed spacecraft. Since we are talking about a spacecraft, however, we have to assume that it would provide its own oxygen to fuel the flames. If the spaceship was manned by a crew, they would be breathing air and this would provide the oxygen needed for a fireball. Oxygen might also be present in some form in the materials used in the spaceship—its metal hull, the fuel used to power it or in compressed air tanks used to provide "squirts" of gas to control the ship's trajectory. When the spaceship went bang, however, we wouldn't see flames appearing the way they do here on Earth.

On Earth a flame licks upwards from material that is burning as it follows the heated rising gases escaping from the combustible material. In space, there is no gravity; there is no "up." Flames would, therefore, spread out in all directions to form a ball of fire. This ball would spread outwards extremely quickly as the force of the explosion sent the oxygen from the ship and any combustible material hurtling out into space. The flame is also likely to snuff out very quickly as all of the oxygen is dispersed and used up. It might seem a bit like a light being quickly turned on and off.

Dangerous debris

Despite the fact that there isn't enough matter in space to support sound waves, it is believed there is enough to support a shock wave, so the spaceship from which we are spectating could well be buffeted by the blast. We would be in far greater danger, however, from the debris sent flying out from the doomed spaceship.

When a bang happens on Earth, if the subject of the explosion is on the ground, debris will be thrown up into the air. It will then fall back down once gravity has exerted its force and the friction caused as the debris forces its way through the densely packed particles of air has slowed it down. The debris from an explosion in space, however, is free as a bird. With no gravity to pull on it and no air to resist it, a piece of space wreckage will carry on traveling at the speed at which it was propelled from the site of the explosion until it eventually runs into something. That could be the gravitational field of a large planet, in which case it will fall to the surface of that planet, or it could be our spaceship, in which case it will slam into us like a guided missile and we're in trouble! Our hull will be peppered with deadly debris! For goodness sake, pilot! Get us out of here!

DRUM BEATING

MODERN COMMUNICATION is extremely sophisticated with its web cams, video phones and satellite links, which allow us to talk to people and to see who we are talking to, even if that person is on the other side of the planet! Centuries ago, however, long-distance communication was delivered with a bang.

The first drums

When early men and women first walked on the face of the earth in Africa, cell phones would have been really useful accessories. They could have telephoned each other to pass on useful hints about where they had found good food or fresh water, they could have kept their hunting parties in touch with one another or warned each other of danger. No doubt they would have found each other's ring tones just as annoying as we do today, too!

Instead of cell phones, the method of long-distance communication developed by early man was drumming. No one knows exactly how drums were invented, but it's a fair bet that someone realized that banging on a hollow log made a noise that could be heard all the way up a valley. From there it would have been a short step to experimenting with tying an animal hide, which would also have made a noise as it was beaten and scraped when it was being cleaned and dried, over one end of the log. Hey presto!—you have a drum.

Drums were adopted by every civilization for communicating, with different rhythms representing different messages. They were used on battlefields and to accompany storytelling, they became an integral part of music and dance as well as religious ceremonies and ritual events. The bang of a drum became an integral part of everyday life. So how does it work?

Several different sizes of drum, producing different types of bangs, are used in a modern "jazz" drum kit

How do drums work?

The answer, as befits the most simple of all methods of communication, is really quite simple itself. Sound is produced by vibration. When the vibration travels through the air as a sound energy wave, we can detect it using our ears.

The striking of a drum skin causes the skin and the body of the drum, to vibrate. The harder you hit the drum skin, the more it will vibrate. This vibration is passed onto the surrounding air, causing the molecules in the air to vibrate and a shock wave of vibration radiates out from the source until it reaches our ears. There, the vibration hits the membrane in the ear called—not by coincidence—the ear-drum. This then vibrates, too, passing the vibration signal on to the brain.

The type of bang produced by a drum can be varied according to how tightly the drum skin is stretched, the size of the drum and the depth of the drum's body. All of these factors will produce differences in the vibration and, therefore, differences in the type of bang we hear. In the hands of a skilled drummer, a modern drum kit with its different sizes of drum will produce bangs that not only make your ears vibrate, they'll set your feet tapping, too!

Gladys had warned Norman countless times about hiding in the freezer

FEELING THE HEAT

EVER BEEN CAUGHT OUTDOORS IN THE COLD when you've been standing still—perhaps waiting for a bus—and you've forgotten your gloves? Chances are that you will, at some point, have rubbed your hands vigorously together and then stuffed them in your pockets. Rubbing your hands together keeps your blood circulating nicely and the friction your hands inflict on each other creates heat—just as fire is made by rubbing two sticks together. What has this got to do with bangs? Energy, that's what.

Everything is energy

Energy is the thing that makes everything happen. You could even say that everything is energy. All matter is energy—the table in front of me, the book you are reading, you and I. Everything can be viewed as a form of energy. On a molecular level, the tiniest particle building blocks that go to make up everything around us cling together because they are charged with the energy to do so. And you know what else? That makes us all indestructible because one of the golden rules of science is that you cannot destroy energy!

OK, maybe we're not quite indestructible in a Superman sort of way. Energy cannot be destroyed, but it can be transferred or converted into another form of energy. We do it every day. You rub your hands together and you turn the kinetic energy from the movement to heat and sound waves. The heat and sound waves then go off and turn into other things. And where did you get the energy from to rub your hands together in the first place? You burn fuel, of course. The energy from the sun that helped to grow the wheat; the energy that was used to turn the wheat to flour; the energy that was used to bake the flour into bread is all stored, in part, in that sandwich you ate for lunch. Your body then breaks down that sandwich to use it as a source of energy to keep you on the go, giving your muscles the energy they need to rub your hands together.

Heat transfer—radiation, conduction and convection

This may seem to be over simplifying (or perhaps over confusing) things, but the point is that energy is never really destroyed, only transferred or converted into another form. The most common byproduct of any energy transference is heat. This means that whenever there is a bang of any description heat will manifest itself in some form or other. We should be thankful for that fact, really, because without it we would not be able to survive. The source of all life on Earth is the sun and the heat provided by this gigantic nuclear reactor in the sky travels through the vacuum of space in the form of radiation, heading towards us at the speed of light. Heat traveling in this manner can be expected to act very much like light, too. It can be reflected, refracted or even absorbed.

Once absorbed, heat can then switch to another mode of transport—conduction. Some materials are better heat conductors than others and metal is

The energy you use keeping yourself warm on a cold day is converted from the energy stored in that sandwich you had for lunch

one of the best. If you stick one end of a metal bar in a fire, for example, it will heat up and the heat will be transferred by conduction to the end that is not actually in the fire.

Convection is the third way in which heat can travel. Convection allows heat to move through a liquid or a gas, such as when the air in a room is heated by a fire or a radiator.

A radiator, in fact, ably demonstrates all of heat's properties of mobility. The hot water or oil in a domestic radiator heats the inner metal surface of the radiator. The heat then travels by conduction to the outer surface of the radiator where it is both radiated into the room and warms the air, which then rises in a convection current, spreading the heat still further.

A destructive force

Heat, then, is a pretty useful form of energy and man has harnessed it to cook food, make tools, generate electricity in power plants and create all of the things we take for granted every day. Heat, however, is also one of the most destructive forces known to man. Apply enough heat to an object and the energy input will cause it to begin to change form. Ice, for example, becomes water

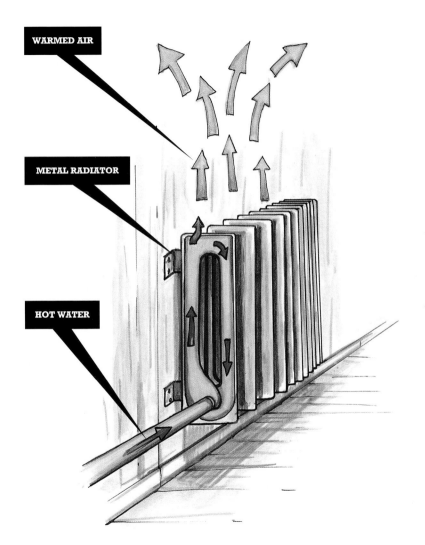

WARMED AIR

METAL RADIATOR

HOT WATER

The creation of the firestorms in Dresden and Hamburg

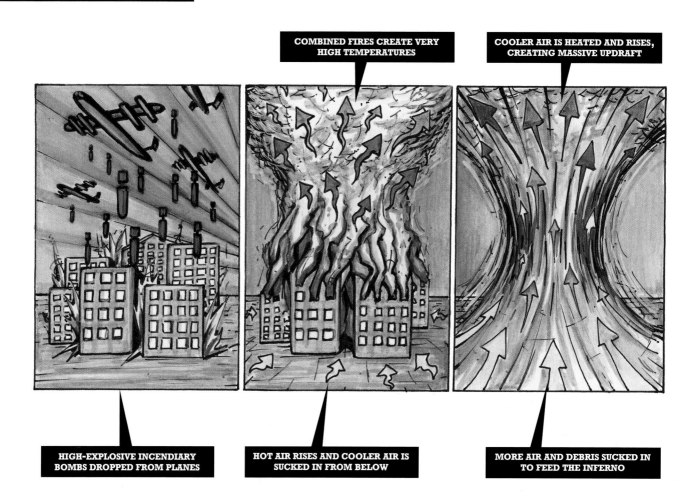

COMBINED FIRES CREATE VERY HIGH TEMPERATURES

COOLER AIR IS HEATED AND RISES, CREATING MASSIVE UPDRAFT

HIGH-EXPLOSIVE INCENDIARY BOMBS DROPPED FROM PLANES

HOT AIR RISES AND COOLER AIR IS SUCKED IN FROM BELOW

MORE AIR AND DEBRIS SUCKED IN TO FEED THE INFERNO

and water becomes steam—a solid changes form to a liquid, which then changes form to become a gas. Apply heat to other objects and their change of form can involve a byproduct of the chemical reaction—fire.

When wood is heated, for example, the cellulose from which the wood is composed will start to decompose. At about 300°F (150°C) the decomposing material becomes smoke, a mixture of hydrogen, carbon and oxygen compounds. When these gases become hot enough, they change form again, combining with oxygen and, in the process, producing flames. By now a chain reaction is under way with more heat being produced by the matter changing form and, therefore, more heat being supplied to encourage the wood to carry on burning. It will continue to burn until all of the cellulose fuel or all of the available oxygen has been used up.

In a big bang, such as when an explosive is detonated, massive amounts of heat are generated in an instant as the chemical reaction of the explosion is triggered. Nitroglycerin produces a temperature of about 9000°F (5000°C) when it explodes—almost as hot as the surface of the sun. The heat produced causes ever greater expansion of the gases produced by the explosion, lending even more destructive power to the blast wave. Like the blast wave, the heat issued from a big bang will dissipate over a distance as the energy is absorbed by other matter and converted to other forms. Near the seat of the explosion, of course, the absorption or conversion can involve the instant incineration of anything within range. This can have a catastrophic effect, even creating what have become known as "firestorms."

Firestorms

During World War II the German cities of Hamburg and Dresden were bombed by the Allied air forces in such a way that firestorms were created. In one raid on Hamburg, over 700 aircraft dropped their incendiary bombs—special high-explosive bombs utilizing materials that burn at extremely high temperatures such as magnesium, phosphorus or petroleum jelly—in a concentrated area, starting hundreds of fires. The fires quickly began to burn out of control and many of them combined to create massive blazes. The city became a mass of flame covering more than 12 square miles (31 km²) and the fires heated the air above the city to temperatures estimated at about 1800°F (1000°C). This caused the air to rise rapidly and cold air was then sucked in at ground level. The force of the air being sucked in, especially along the streets and alleys of a city, created a tornado-like wind traveling at 150 mph (240 kmph), which picked up all sorts of debris and then deposited it in the flames to continue feeding the fire with both fuel and oxygen. People, too, were hurled into the flames. Others caught in the maelstrom suffocated as the fire used up all the available oxygen and warmed the rest to send it skyward.

The firestorm effect is something of which, on a far smaller scale, firefighters are very wary in domestic blazes. The rapid movement of air can make a fire extremely difficult to control.

The destructive power of heat, then, can be quite awesome, but without it we'd all be left rubbing our hands together like crazy in cold weather!

BOTTLED BANGS

THE POPPING OF A **C**HAMPAGNE **C**ORK is a universal sound of celebration, a noise that is guaranteed to start a party with a bang—surely one of the most welcome explosions ever heard!

Explosions? Are these good-time bangs really explosions? Of course! Almost all big bangs rely on expanding gas to lend them their power and a popping Champagne cork is no different. The expanding gas is carbon dioxide and the power it supplies can propel a Champagne cork out of its bottle at a speed of up to 60 mph (97 kmph)! Look at a Champagne bottle and you will see that it is made from substantially thicker glass than a normal wine bottle. This is to contain the pressure created by the gas that is trapped inside along with the wine.

So where does the gas come from? In Champagne, just as in beer, the carbon dioxide is produced by the fermentation process.

Making bubbles

Yeast, a living organism, is the most important part of the fermentation process. As it reproduces, it consumes the sugar that has also been added to the solution. The two byproducts of yeast's reproduction process are alcohol and carbon dioxide. This process is skillfully controlled by wine makers and brewers who need to know when their product has just the right amount of fizz and just the right amount of alcohol!

Just the right amount of fizz for Champagne is more than merely that initial release of gas. The noise created when the gas rapidly expands out of the bottle creating a shock wave that we hear as a bang is only the first explosion. Gas continues to escape from the Champagne even after it has been poured into a glass. The wine fizzes as bubbles of carbon dioxide liberate themselves from the liquid and rise to the surface. Those bubbles

There can be as many as 45 million mini bangs in a bottle of Champagne

POP

then pop, releasing their little parcels of gas into the atmosphere. Each of those pops is actually a mini explosion lasting only the tiniest fraction of a second and they come in such rapid succession that they combine to make a fizzing sound. If you could record that sound and slow it down enough you would be able to hear a series of bangs—a big series of bangs. It has been estimated that a bottle of Champagne contains as many as 45 million mini explosions!

Fizzy pop

The same principle applies to other fizzy drinks such as soda or carbonated mineral water. Mineral water, in fact, is the oldest fizzy drink of all, with the Ancient Greeks and Romans having established baths in places where the rock structure was such that naturally carbonated mineral water literally bubbled up out of the ground. Bathing in the stuff was deemed extremely therapeutic. It wasn't until the late 18th century that an English chemist discovered how to make artificially carbonated water and before long all sorts of fruit juices and syrups were being added to produce the sorts of flavored sodas we have today.

Champagne is still the one that goes off with the biggest bang, though, so no one can really say: "I get no kick from Champagne..."

THE MONT BLANC TRAGEDY

ON THE MORNING OF THURSDAY DECEMBER 6 1917, as World War I raged in Europe, the French cargo ship *Mont Blanc* was making its way to Halifax, Nova Scotia, where it was to await a suitable convoy to escort it across the Atlantic into the war zone.

The port area of Halifax was extremely busy with ships carrying troops, supplies and munitions and the *Mont Blanc* was herself a munitions ship. In her holds and on her decks she was carrying 300 rounds of ammunition for her own two deck guns; 10 tons (10 tonnes) of gun cotton, the explosive used as a propellant for artillery shells; 35 tons (36 tonnes) of Benzol, a type of fuel; 2,300 tons (2,337 tonnes) of picric acid, a chemical used in explosives; and 200 tons (203 tonnes) of TNT high explosive. The *Mont Blanc* was the biggest floating bomb ever to set sail.

In the narrow channel approaching Halifax Harbor there was a strict rule that ships heading into and out of the docks passed each other "port to port," meaning that the ships passed each other's left-hand side. As the *Mont Blanc* negotiated the narrows, however, a great confusion arose between the *Mont Blanc* and a Norwegian ship, the *Imo*. The larger *Imo* rammed the *Mont Blanc*, cutting a gash in the *Mont Blanc*'s hull and causing sparks, which started a fire directly beneath the drums of Benzol stowed on deck.

Realizing the danger, the *Mont Blanc* crew abandoned ship, and the blaze raged out of control as the ship drifted slowly towards Halifax. Within 20 minutes it had drawn crowds of fascinated onlookers. Had the *Mont Blanc* been flying the regulation red flag to warn that she was carrying explosives, people might well have kept their distance. The burning ship set Halifax pier alight and, as the local firefighters started to fight the blaze, the *Mont Blanc* exploded at 9.05 a.m. with what is thought to be the biggest non-nuclear man-made bang ever.

There was a blinding white flash and a massive blast, which devastated 325 acres (132 ha) around the port. The nine Halifax firefighters were among the 1,600 people who were killed instantly, although the death toll would later rise to 2,000. There were 9,000 serious injuries. The blast was reported to have shattered windows 50 miles (80 km) away and to have been felt over 200 miles (322 km) away. The 3,000-ton (3,050-tonne) *Mont Blanc* was blown to pieces, the barrel of one of its deck guns was found over 3 miles (5 km) away and part of her anchor, weighing ½ ton (0.5 tonne), was discovered 2 miles (3 km) from the explosion.

For the people of Halifax, this was a tragedy never to be forgotten and, among the many memorials to those who died on that day, the Halifax Fire Department erected a monument outside Fire Station 6 to commemorate the men who died trying to fight the fire on the *Mont Blanc*.

The explosion of the *Mont Blanc* cargo ship in 1917 resulted in thousands of casualties and devastated a vast area of Halifax, Nova Scotia

Rockets rely on Isaac Newton's third law of motion and, although they are more accurate nowadays, they work on the same principle established by the Chinese a thousand years ago

ROCKETS

IT SUFFERS FROM EXPANDING GAS, it has a thunderous roaring noise coming from its bottom and it moves absurdly fast. Yes, beer and beans can do that to you, but I'm talking about rockets. Rockets really do go with a bang. They take off with a bang and, given that so many rockets have been designed to deliver an explosive payload, they can also land with a pretty impressive bang!

The principles of rocketry

One of the things that has always made space travel such a risky business is that the spacecraft used by astronauts depend upon explosive rocket power to take them into space. Basically, they still rely on the same principles of rocketry that were first established by the Chinese a thousand years ago.

Remember the Chinese cook who invented gunpowder back on page 59? That black powder he concocted became all the rage. You were nobody in 11th-century China unless you could pull off a few impressive flash-bang stunts with the black stuff. Following the invention of the first artillery, experiments with gunpowder soon led to development of the rocket. A rocket, after all, uses the same principle as a cannon (see page 94). In fact, turn a cannon round, narrow the opening and set it off and you have pretty much invented a rocket. Not convinced? What's at work here is Isaac Newton's third law of motion, which states that for every action there is an equal and opposite reaction. Cause an explosion in one end of a cannon and the expanding gases will push a cannonball up the barrel and out towards the enemy. At the same time, the cannon will be pushed in the opposite direction. So, why not put some black powder in a tube like a cannon barrel, and see if the force of the expanding gas exiting the open end can propel the whole thing forward? It couldn't have taken the Chinese too long to give it a go.

Rockets as weapons

The recipe for the black powder was adjusted slightly to make it burn longer rather than simply blowing the rocket to pieces and soon rockets were being used as weapons. Rockets attached to arrows increased the range of the latter and could also be used to set fire to straw-roofed buildings or the sails of enemy ships. When a charge of the original, more explosive, black powder mix was carried as a payload, the rocket became an immensely destructive weapon.

It took many years of further development before rockets had the accuracy or the range of conventional artillery, although their use spread from the Far East, through the Indian subcontinent to the Middle East and Europe, following the established trade routes. Rockets drifted in and out of fashion with the military, for many years failing to keep pace with the more rapid advances in artillery technology. This was especially true in Europe, although rockets did cause a sensation wherever they were used.

In the early 19th century, a British army officer, William Congreve, developed a metal-cased rocket with a range of over ½ mile (0.8 km), delivering up to 7 lb (3 kg) of incendiary explosive. Congreve's rockets were used against the United States in 1812 when Fort McHenry in Maryland was bombarded. In *The Star Spangled Banner*, the line about the "rockets' red glare" refers to the British rockets that rained down on Fort McHenry.

Rocket development really took off in the 20th century when German scientists working under Wernher von Braun in World War II developed the V-2 rocket, a missile that could deliver a payload of around 1 ton (1 tonne) of high explosive over a range of 250 miles (400 km). Von Braun and other key members of his team were whisked off to America after the war and played a key role in the U.S. space program.

Since the end of World War II in 1945, rockets, missiles, their guidance systems and the fuels they burn have become ever more sophisticated, but the basic principle remains unchanged. Fire an expanding gas in one direction to propel a rocket in the other. To demonstrate the rocket effect for yourself blow up a party balloon, hold the neck closed then let it go. The gas in the balloon is held under pressure and wants to expand. It does so by escaping through the mouth of the balloon. For a more spectacular effect, build your own rocket using household items, as described overleaf.

ISAAC NEWTON

1642—1727

BORN IN LINCOLNSHIRE, England, on Christmas Day in 1642, Isaac Newton's birthday now falls on January 4 following a change of dates after the Gregorian calendar was adopted in Britain in 1752.

Baby Isaac was born in a grand manor house; his father was a wealthy land owner who died a few months before his son's birth. Following a troubled childhood, much of which was spent away from home in the custody of his grandmother or other carers after his mother remarried, Newton overcame what had been regarded during his early school days as academic ineptitude to study at Cambridge University. Although he started out as a law student, Newton soon developed a passion for science. When Cambridge University closed down for two years due to the plague in 1665, Newton returned home to Lincolnshire where he worked on developing revolutionary theories in astronomy, physics and mathematics.

He showed that white light was actually a combination of different colors of light, developed theories about how gravity worked and showed how objects reacted to the application of force. By 1687, when he published *Philosophiae Naturalis Principia Mathematica*, acclaimed as the foremost science book of all time, Newton was widely regarded as one of the world's greatest scientists.

In 1696, Newton became head of the Royal Mint, in charge of the country's currency, and in 1705, at the age of 63, he was knighted by Queen Anne, the first scientist ever to receive such an honor. Sir Isaac Newton died in 1727.

MAKING A ROCKET

You will need:

A SMALL SHEET OF CARD (A COUPLE OF OLD
  POSTCARDS MIGHT DO)
STICKY TAPE
A LARGE EMPTY PLASTIC DRINKS BOTTLE
WATER
A NAIL
A CORK THAT WILL FIT THE BOTTLE TIGHTLY
A NEEDLE ADAPTER USED FOR INFLATING
  BASKETBALLS
A BICYCLE PUMP

1 **Cut four "fin" shapes from the card. Your
bottle rocket, half-filled with water, will
stand on them, so they have to be big
enough and strong enough to hold it a few
inches off the ground. Cut a small slit
halfway along the base of each fin, then
fold back two little flaps, one in each
direction, that you can use to tape the fin**

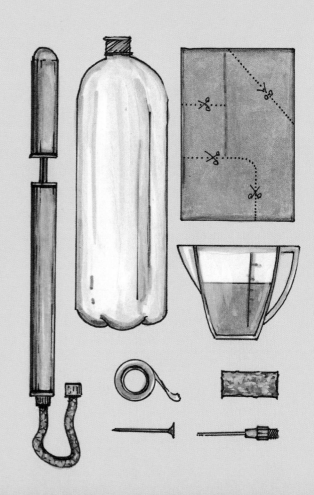

to the neck end of the bottle to create the rocket stand. Once your fins are in place, half-fill the bottle with water.

2 Use the nail to make a hole in the cork then push the cork into the neck of the bottle. Insert the needle adapter into the nail hole, pushing it far enough in so that you will be able to pump air into the bottle. Attach your pump to the adapter.

3 Go outside, find a safe open space where you won't annoy anyone, stand your rocket on its fins and start pumping.

4 Keep pumping until the air pressure inside the bottle blasts the cork out and your rocket takes off with a bang—and a shower of water!

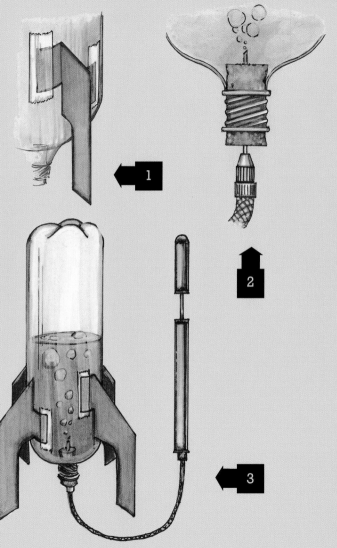

JET ENGINES

WHAT HAVE HAIR DRYERS, vacuum cleaners and jet engines got to do with bangs? Hair dryers are not usually supposed to go bang are they? You certainly wouldn't want to be pointing one at your head when it did! What about vacuum cleaners? Surely they go bang only when some idiot vacuums something he shouldn't! Jet engines? Well, you wouldn't want to be flying in an airplane when its engines went bang, would you? Actually, yes, you would. In fact, you wouldn't want to be flying in a plane when its engines didn't go bang. Running a jet engine, you see, is like feeding one long, continuous explosion. OK, so what's that got to do with hair dryers and vacuum cleaners?

Invention of the jet

The jet engine, like so many other marvelous inventions, might never have come about had it not been for war—in this case World War II. Although various inventors and engineers had experimented with jet or gas turbine engines before, it wasn't until the late 1930s that the technology and the materials existed to build a practical jet engine that would produce enough power to keep an aircraft aloft. The first person to design and build such an engine was Sir Frank Whittle, a British Royal Air Force officer who successfully "bench tested" his design in April 1937. Six months later, a German engineer called Hans von Ohain also tested a jet engine. The two designs were similar but there was neither collusion nor cooperation between the two men. Germany was under the rule of the Nazis and all of Europe was bracing itself for war.

Because of the impending war, Whittle found it difficult to solicit government backing for his project. British resources were being plowed into the manufacture of less risky, tried and proven, more

conventional aircraft manufacture. In Germany, however, von Ohain had the backing of aircraft manufacturer Ernst Heinkel and on August 27 1939, Flight Captain Erich Warsitz became the first man to fly a jet aircraft when the Heinkel He 178 test plane took to the air. It would be almost two years before Whittle's engine and test aircraft were ready to fly but there was no great international race to develop the first operational jet fighter aircraft. Conventional aircraft were in plentiful supply on the Allied side and the Allied commanders wanted aircraft on which they knew they could rely. The only real race was happening within Germany where the jet aircraft on which they were ultimately concentrating—the Messerschmitt Me 262—had shown it could achieve speeds up to 150 mph (240 kmph) faster than the fastest of the Allied conventional fighter planes. It was seen as a weapon that could save Germany from defeat. In the end, Nazi leader Adolf Hitler's poor judgment coupled with Allied efforts to destroy production of the Me 262 caused delays which meant that both the British and the Germans had jets entering operational service around the same time in the spring of 1944.

A hair dryer sucks in air and blows it out again, just like a vacuum cleaner, but you wouldn't want to use one to clean up your house!

Suck and blow

Since then, jet engine and jet aircraft design have undergone constant development with the fastest modern jet fighters now achieving speeds around three times faster than that of the Me 262 of 60 years ago. Nevertheless, the modern jet engine still works in much the same way, which is where the everyday hair dryer and vacuum cleaner come in.

Hair dryers and vacuum cleaners both suck in air and blow it out again. In the case of the hair dryer it is blown out in a jet that dries your hair. A vacuum cleaner sucks in dust along with the air then filters out the dust (normally into some kind of bag) before the air is expelled. A jet engine also sucks in air. It then pushes the air out of its exhaust port with enough force to send an aircraft hurtling through the sky. The jet generates this force with the help of a really big bang—a continuous controlled explosion. When the air is sucked into the engine it is first compressed, forced into a

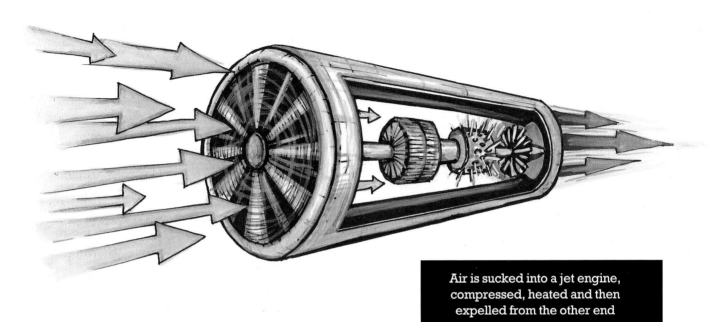

Air is sucked into a jet engine, compressed, heated and then expelled from the other end

small space under pressure. Fuel is then injected into the compressed air and ignited. The expanding gases from the explosion then shoot out of the tail pipe. Old Isaac Newton's (see page 83) third law of motion then comes into play—for every action there is an equal and opposite reaction. The expanding gases exiting the jet's exhaust push the aircraft forward. To provide even more thrust, the exhaust gases from a jet are sometimes also fed through an afterburner where more fuel is injected and ignited to produce a secondary explosive thrust.

Jets versus rockets

The fact that a jet engine needs to suck in air means that it can be used only inside Earth's atmosphere where there is air, or rather oxygen, to feed the burning of the fuel. That's why rockets, which carry their own oxygen supply either within the fuel or in a separate tank to add to the fuel, are used for space travel (see page 80) and jets are not. Quite simply, in the vacuum of space there is nothing for the jet engine to suck in and burn, so it won't work. And neither would your hair dryer or vacuum cleaner!

LISTEN UP!

BANG! DID YOU HEAR THAT? No, of course you didn't. That was just a word written on a page. You can't hear a word written on a page. Now read it aloud. "*BANG!*" That's better. Now you can hear it, and so can everyone else in the room. If they'd paid good money to hear a proper bang, however, they'd be pretty upset. "That wasn't a real bang," they'd say. "That was just you impersonating a bang." But they wouldn't know what they were talking about, because what you provided them with was every bit as real as any other bang.

Getting an earful

We've already learned how, in an explosion, the particles that make up the explosive or combustible material become extremely agitated as they change form to become gases. The extremely agitated gas molecules then expand rapidly, passing on their energy to agitate, vibrate or push everything around them. This includes the molecules of air in the atmosphere.

The vibration created at the site of an explosion and passed on by the expanding gas or anything else that the gas has caused to vibrate travels through the air in a wave form. When the air around our heads picks up the vibration, we can detect it. The highly sophisticated sensors that we use to detect this vibration are our ears and the first parts of our ears to see (hear!) any action are the pinna, the wobbly fleshy bits sticking out from each side of your head. Without the pinna we'd be practically blind. No, I don't mean you couldn't keep your glasses on and your hat would fall down over your eyes. We would be almost blind to sound because the pinna and the auditory canal of the outer ear act like a funnel, directing vibrations in the air onto the tympanic membrane, or ear-drum. The shape of the pinna also helps you to determine from which direction the vibrations are coming—where all the noise is coming from.

How the ear-drum works

The ear-drum is at the end of the ear canal and separates the outer ear from the middle ear. When it is hit by a sound wave, it vibrates, just like the skin of a drum. It is so incredibly sensitive that even the slightest changes in air

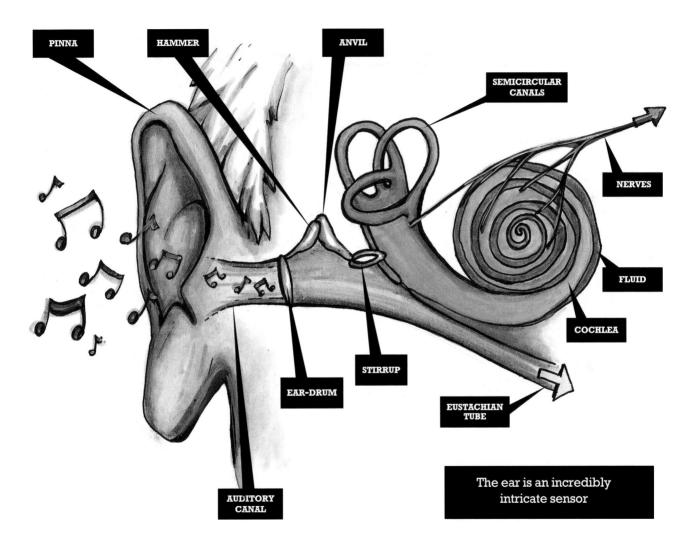

PINNA

HAMMER

ANVIL

SEMICIRCULAR CANALS

NERVES

FLUID

COCHLEA

EAR-DRUM

STIRRUP

EUSTACHIAN TUBE

AUDITORY CANAL

The ear is an incredibly intricate sensor

pressure can be detected, allowing it to pick up the tiniest vibrations of sound. Higher pitched sound waves will cause the ear-drum to vibrate back and forth more quickly than lower pitched sound, while louder sound (more powerful vibrations) will make it move further as it vibrates. Special muscles control how taut the ear-drum is pulled, also operating a kind of fail-safe system that will attempt to cut out some of the vibration if really loud noises threaten to cause so much rattling that the ear might be damaged. So the ear-drum is the main sensory element of the ear. The other parts of the ear work together to translate and pass on the signals it receives.

Given the sensitivity of the instrument we are dealing with here, it will come as no surprise that the ear-drum passes on its vibrations via the smallest bones in your body, known as the hammer, anvil and stirrup bone. These three bones work together in a mechanical sort of way to convert the vibrations passed on by the ear-drum into a kind of piston motion in the stirrup bone. This pushes on another membrane known as the oval window, which passes on the sound wave from the middle ear to the fluid-filled inner ear. The tubes in the inner ear form three loops known as semicircular canals, which sense changes in the position of our heads, and also the cochlea, which forms the next link in the information chain.

The cochlea

The cochlea is made up of three tubes coiled in a spiral shape. These tubes are composed of a range of delicate fibers, some short and stiff, some longer and more flexible. As the oval window responds to the agitation of the stirrup bone, it sends pulses through the fluid in the cochlea. These represent the frequency of the sound wave and travel the length of the cochlea until they find the fibers that suit them best. There are around 30,000 fibers to choose from and each will vibrate only in response to a sound wave of a certain frequency. They then resonate at exactly the same frequency. High-frequency waves will vibrate the fibers nearer the oval window while low-frequency waves will travel past these to be "caught" by the more flexible fibers further into the cochlea.

As the chain reaction continues, the fibers pass the vibration to a range of thousands of hair cells and it is the movement in these cells that stimulates the cochlear nerve. The nerve finishes the job of converting the vibration to a nerve impulse, an electrical signal that it sends to the brain for decoding. Your brain then tells you that you just heard a bang.

Auditory signals to the brain

The brain can identify and classify millions of sounds of all sorts in an instant. It can identify the engine noise, tire noise, exhaust note and general air disturbance of a passing car; the words, guitar chords and drum beat of a song playing on that car's radio; a bird twittering in a tree; a dog barking; a phone ringing; a leaf scraping along the sidewalk or even your own footsteps. Noisy in your neighborhood, isn't it? All of these sounds are picked up as different vibrations in the air.

In order to speak, we vibrate our vocal chords to make noises, a skill that we develop as we mature. Our

BANG!

brain controls that, too. Somewhere in the brain's memory banks it stores the information that allows us to recognize different sounds. It then draws on that information to allow us to recreate different sounds. In speech we form these sounds into words.

When you shout "*BANG!*" you are using your memory bank to help you recreate the sound that we recognize as coming from some kind of explosion. You do this by vibrating your vocal chords—you create a vibration, or sound wave, in the air. That's how the other people in the room managed to hear you—they picked you up with their pinna.

Those other people, though, claimed that you only impersonated a bang. But that wasn't just an impersonation, was it? You set up a vibration in the air. The sound of an explosion is a vibration in the air, too. In other words, they are exactly the same thing. The sound of a bang is the sound of a bang, whatever the source!

YOU'RE FIRED!

THE ONE PIECE OF MODERN TECHNOLOGY that has remained in use, virtually unchanged in principle since its inception, is the cannon, or artillery piece—the big gun that produces a mighty big bang.

Just as gunpowder was invented in the Far East, so too were firearms, and a firearm is probably a pretty accurate description of the first cannon. Known as hand cannons, they were simple bronze tubes that could fit in the palm of your hand, which were filled with powder and used to fire a metal or stone ball. When held at arm's length it would look like "fire" was shooting out of your arm. Previously, bamboo may have been used for rocket casings, but it would never have been strong enough to withstand the explosive power of an artillery charge going off. Neither, sadly, were some of the early cannons. Artillery pieces were often as dangerous for the user as they were for the intended target as the charge could just as easily blow the cannon apart as it hurled the cannonball out of the barrel.

Early European cannons

Experimentation with artillery and the development of the cannon was happening in the Far and Middle East long before cannons first appeared on the battlefields of Europe, but it was the superior foundries and metal-casting techniques of the Europeans that made the cannon an indispensable battlefield weapon. Having acquired the infamous black powder recipe from the Chinese (see page 59), the Europeans lost little time in putting it to use as a weapon. One of the earliest illustra-tions of a cannon in use dates from around 1325 and shows an early type of cannon firing a huge arrow at the walls of the French port of La Rochelle. It is generally accepted that the first reliably documented use of cannons on the battlefield relates to the Battle of Crécy in 1346 when the English King Edward III deployed three cannons against the French.

Despite the fact that the Church denounced the use of black powder, maintaining that only sorcerers in league with the Devil would dabble in this new black art, once the destructive force of the cannon had been seen and felt, there was no going back. Although English archers with their longbows were a formidable force, they could not match the range of the artillery pieces being introduced by the mid-15th century. A longbow had an accurate range of up to 200 yards (183 m). A cannon could hurl its ball anything up to 1 mile (1.6 km).

Cannons were expensive to produce but as national armies with national treasury funding began to replace the private forces previously raised by noblemen, more money was available for the purchase and development of cannons. By 1600 a number of different types of artillery pieces, from mortars to more traditional cannons, were in use and the basic method of operation that was to remain unaltered for centuries had evolved.

How cannons work

In its crudest form, the cannon was a long metal tube open at one end and sealed at the other with the exception of a tiny hole through which to introduce the fuse or ignition powder. You poured some black powder, the propellant charge, into the barrel and rammed it all the way down to pack it into the sealed end. Then you pushed your cannonball, stones, chains or whatever other nasty, vicious stuff you were going to fire at your enemy, down the barrel and rammed it up against the powder charge. Through the tiny hole at the sealed end you then inserted a fuse or poured some more gun-powder. Finally, you lit the fuse or igniter powder and set off the propellant charge. The detonation of the main charge produced the rapidly expanding gases of an explosion which, in order to escape from the confined space of the cannon, pushed the cannonball up the barrel at a great rate. The cannonball then exited the end of the barrel in the general direction of the enemy.

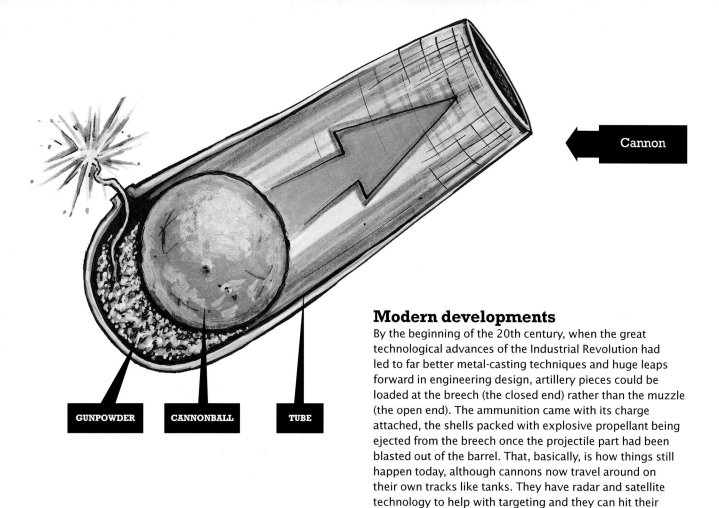

Cannon

GUNPOWDER

CANNONBALL

TUBE

Modern developments

By the beginning of the 20th century, when the great technological advances of the Industrial Revolution had led to far better metal-casting techniques and huge leaps forward in engineering design, artillery pieces could be loaded at the breech (the closed end) rather than the muzzle (the open end). The ammunition came with its charge attached, the shells packed with explosive propellant being ejected from the breech once the projectile part had been blasted out of the barrel. That, basically, is how things still happen today, although cannons now travel around on their own tracks like tanks. They have radar and satellite technology to help with targeting and they can hit their targets from beyond the horizon with accurate ranges easily exceeding 20 miles (32 km).

Twentieth-century technology vastly improved the manufacture, reliability, range, and accuracy of the cannon

SMASHING THE SOUND BARRIER

CHARLES ELWOOD YAEGER is undoubtedly a remarkable man. Born in Myra, West Virginia, in 1923, he enlisted in the U.S. Army Air Corps as soon as he graduated from high school. Before long he was flying fighter planes over France in World War II. He managed to shoot down one German fighter plane and one bomber before he himself was shot down. Evading capture with the help of the local French Resistance, he made his way back to England via Spain and went on to fly a further 56 combat missions, shooting down another 11 enemy aircraft.

As a fighter ace and war hero, you might think that Chuck had already claimed his place in history, but his most notable achievement was yet to come. On returning to the United States in 1945, the former corporal-now-captain became the project officer on the Bell X-1, America's highly advanced rocket research aircraft.

On October 14 1947 Yaeger guaranteed his place in history by becoming the first man officially to fly faster than the speed of sound, taking the Bell X-1 through the sound barrier and producing in the process a unique bang—the world's first sonic boom.

What is a sonic boom?

Actually, although the incredible Yaeger may have been the first man to go supersonic, he most definitely did not produce the first sonic boom. A bullet can do that. So, too, when it is cracked expertly, does the tip of a whip.

Exactly how is this high-speed bang produced? Well, the bang isn't high speed, it is a sound wave that moves at the speed of sound. The problem is, that's just not fast enough to outrun some things.

Imagine a pebble dropped in a pond. It produces a circular wave emanating from its impact with the surface. That's pretty much how a sound wave emanates from a source vibration. Now imagine a speedboat on the pond. Dropped into the water it would produce a circular wave, but moving across the pond it catches up with that circular wave and crashes through it, setting up a bow wave, or wake. An aircraft hurtling through the sky produces a great deal of noise. This noise spreads, like the circular wave of the pebble, in all directions—including ahead of the plane—at the speed of sound. When an aircraft flies faster than the speed of sound, it crashes through its own sound wave, setting up a shock wave much like the boat's wake. This spreads behind it not like the "V" of the boat's wake, but in all directions through the air, like a cone. When the cone sound energy shock wave reaches our ears, we hear it as a bang—the sonic boom.

And Chuck Yaeger from Myra? He went on to become General Yaeger and, at the age of 74 he celebrated a brilliant military career by breaking the sound barrier again on the 50th anniversary of his first supersonic flight, this time in an F-15 fighter plane.

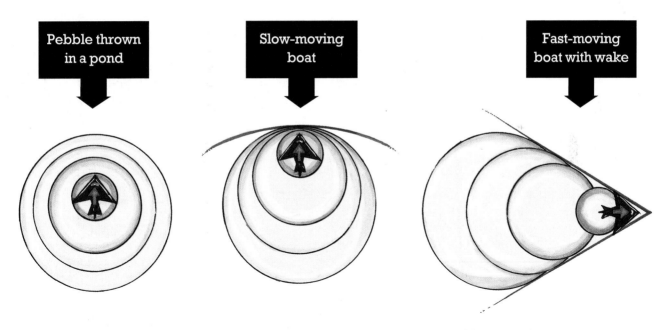

Pebble thrown in a pond

Slow-moving boat

Fast-moving boat with wake

UNDERWATER BANGS

WE'VE ALREADY LOOKED AT WHAT HAPPENS when a bang is detonated in outer space, but what about a bang that goes off in what you might consider to be the opposite of outer space—the depths of the ocean? Unlike in outer space which we can regard as "empty"—a vacuum—underwater is far from empty. It's full of, well... water. The deeper you go, of course, the more water there is above and around you. The weight of that water pressing down on you would, eventually, be enough to crush you. In fact, even the strongest, most modern submarines risk being crushed by the water pressure if they dive too deep.

Under pressure

Water pressure has a major effect on the properties of a big bang. Unlike in outer space, you would still hear the bang as sound waves can travel rather well through water. We don't hear so well underwater because the water pressure on our ear-drums means they don't pick up the vibrations so well, but the sound of the bang is happening none the less.

Similar to outer space, there is no readily available oxygen to burn underwater, so once any oxygen released by the combustible material or explosive has been consumed, the flash will be extinguished.

The main effect of the water pressure, however, is to contain the expanding gases, which must work much harder against the water than they have to do in the open air. The water pressure contains the gases in a bubble, or series of bubbles which billow in and out as the water pressure fights the gas pressure and the bubble of gas, being lighter than water, rises to the surface to escape into the atmosphere.

The same is, therefore, true of the shock or pressure wave. It tends to be retarded by the pressure of the water and, if the explosion is close enough to the surface, much of the normally destructive energy of the blast can be dissipated harmlessly, throwing water into the air.

The Dam Busters

This was both a problem and a solution for British inventor Barnes Wallis (see page 46) during World War II. The British had tried unsuccessfully to destroy the massive dams in the Ruhr Valley, which provided hydro-electric power to German industry. Existing bombs weren't big enough to destroy the dams, even if a direct hit could be scored. Wallis hit upon the idea of a spinning, bouncing bomb, which, when dropped from a specific height, would skip across the surface of the reservoir behind the dam like a stone. When it hit the dam its

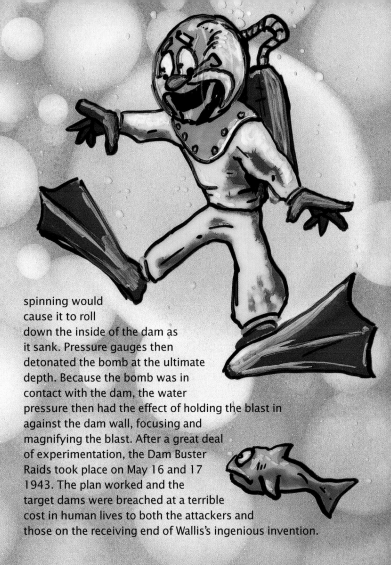

spinning would cause it to roll down the inside of the dam as it sank. Pressure gauges then detonated the bomb at the ultimate depth. Because the bomb was in contact with the dam, the water pressure then had the effect of holding the blast in against the dam wall, focusing and magnifying the blast. After a great deal of experimentation, the Dam Buster Raids took place on May 16 and 17 1943. The plan worked and the target dams were breached at a terrible cost in human lives to both the attackers and those on the receiving end of Wallis's ingenious invention.

VOLCANOES

REMEMBER KRAKATOA? No, I didn't mean you look old enough to remember back to 1883! I mean remember "Krakatoa" back on page 34. That volcanic eruption caused one of the biggest bangs ever heard and if we stick around long enough, it's a sure thing that there will be another one just like it sooner or later. The next big volcanic bang may be some way in the future, but it really started long ago in the past—all the way back, in fact, to the beginning of time.

The composition of Earth

When the first stars reached the end of their lives as stars and blew themselves apart, they hurled hot gas and molten elements all over the universe. These cooled to become planets or carried on burning as smaller suns, smaller nuclear reactors still processing the fallout from the Big Bang (see page 10).

As the space debris that crashed together to form the Earth cooled, a crust formed around the planet, like a scab covering a wound, and it is on this crust that life evolved and that we now live. The crust also provides insulation for the inside of our planet and this protective blanket, along with our atmosphere and the phenomenal pressure on the planet's interior caused by the mass of the Earth pressing in towards its core, ensures that the interior of the planet remains a molten mass, a throwback to days long gone when Mother Earth was still a fiery kid.

The crust on which we live is up to 40 miles (64 km) thick under the continental land masses and only 3—6 miles (5—10 km) thick beneath the oceans. Beneath the crust is a layer called the mantle (actually divided into a number of mantle layers), which is about 1,800 miles (2,900 km) thick and mainly solid but at a temperature of between 1800°F and 5400°F (1000—3000°C). The hotter it gets, the more plastic it becomes, turning to molten rock, or magma.

Below the mantle is a layer of liquid iron and nickel about 1,400 miles (2,250 km) thick and at a temperature of anything up to 7200°F (4000°C). Beneath that is the center of the earth—a core ball of iron and nickel at about 9000°F (5000°C) but under so much pressure that it is probably solid. Nobody's ever actually been there to check!

Mount Vesuvius in Italy has erupted 50 times since it destroyed Pompeii in AD79

Volcanic activity

Because the mantle, inner and outer cores of the earth are molten, they are actually quite flexible, but the cold outer crust is thin and brittle. As the earth spins the crust moves, "floating" on the mantle, but because it isn't as flexible as the mantle, it can crack or rupture. When it does crack, gases trapped in the magma of the mantle, in conjunction with the pressure created by the intense heat, force the magma up through the fractures in the outer crust.

The amount of gas trapped in the magma at that particular spot together with the consistency of the magma determine how violently the magma erupts through the crust. If there is a lot of trapped gas, then the magma will be under more pressure. If the magma is quite fluid it will allow the gas to bubble through towards the surface of the crust more easily and the gas will force less of the magma out through the fracture. If the magma is quite dense, however, it will hamper the progress of the gas and cause a build-up of pressure, which results in the gas taking a lot of the magma with it when it finally escapes to the atmosphere. In other words, there is a big eruption.

Stratovolcanoes are tall, steep-sided mountains whose slopes lead up to a small crater at the summit. These are normally associated with Plinian eruptions, which have produced massive amounts of pyroclastic material—lava, rock, ash, and other debris.

Scoria cone volcanoes are smaller mountains with much wider craters at the top. These are produced by Strombolian eruptions—short but spectacular blasts of fiery molten rock, with little subsequent lava flow.

Shield volcanoes are formed when lava oozes out through the ground with a minimum of fuss, creating a wide, low hill that you might not even realize is a volcano. Don't be fooled, though, because shield volcanoes erupt far more frequently than any other type.

Magma, once it is on the surface of the crust and starting to cool, is called lava.

The types of magma, eruption and fracture in the earth's crust all go towards creating different types of volcanic activity and different types of volcano. Plinian eruptions have more solid magma and a high gas content and produce eruptions such as the one that buried Pompeii in AD79. These eruptions can hurl magma and other debris as high as 30 miles (48 km) or more into the air. Hawaiian eruptions are far less destructive, with low gas content and low-resistance magma producing sluggish lava flows, although there can be the occasional highly impressive fire fountain when spouts of bright orange molten lava froth hundreds of feet in the air.

There are three main shapes of volcano that form on land—stratovolcanoes, scoria cone volcanoes and shield volcanoes.

Areas where the earth's crust is prone to fracturing will develop more volcanoes and, once a volcano has formed, pressure from below will often force other vents or fissures to open in the volcano, causing further eruptions. Whether these eruptions come months, years or centuries apart, the bigger ones produce bangs that literally turn the world inside out, with the magma giving us a glimpse of what the surface of our planet looked like many millions of years ago.

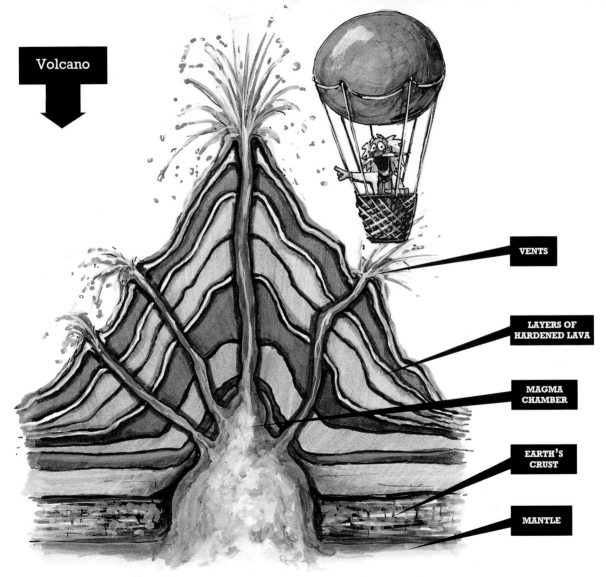

Volcano

VENTS

LAYERS OF
HARDENED LAVA

MAGMA
CHAMBER

EARTH'S
CRUST

MANTLE

It is now widely accepted that the sun's rays can seriously damage human skin, but as long as we use proper protection and don't stay out in the sun for too long, we can all still enjoy a day at the beach

NATURE'S ARSENAL

WHEN IT COMES TO CREATING MAYHEM and general chaos, nobody does it better than Mother Nature. In a competition between man and Mother Nature to produce the most impressive bang, man could play a pretty strong card with the thousands of tons of incendiary bombs he has used in the past to create firestorms that have destroyed entire cities. But Mother Nature could counter with a small volcano. Man could field one of his amazing blast bombs that completely flatten hundreds of square yards of the densest jungle or townscape. But Mother Nature could just whistle up a bit of a tornado or possibly a hurricane.

If, during the course of this hypothetical arms race man was then to reveal his nuclear arsenal, Mother Nature need only slide back a couple of clouds to reveal the sun, the biggest, most powerful nuclear reaction in this neck of the universe. Is there nothing man's got that can beat nature? What about a ray gun? Forget it. Mother

Nature's been shooting rays around since the beginning of time. Some of them we can see, some of them we can't see, which is a shame because it's the ones we can't see that can really do us damage.

Radiating rays

The massive ongoing nuclear reactions of the sun emit electromagnetic radiation across the entire spectrum. Our eyes can detect this radiation when it falls within the wavelengths of the visible light spectrum, which ranges from red and orange through yellow, green and blue to indigo and violet. The different colors of light, as we now know, combine to produce white light. We can't see radiation energy traveling in the higher-energy waves such as X-rays or the lower-energy radio waves. In this case, the old adage of "What you can't see can't harm you" couldn't be more wrong.

Applying energy to an atom causes electrons to expand their orbit around the nucleus. When they drop back, they release a burst of energy

To understand why, we have to look again at how the big bangs of the sun produce radiation and how organic creatures such as ourselves hang together.

First of all it's worth remembering that everything on Earth, from rocks and metal to glass, hair, skin or teeth, is made up of atoms. Everything is composed of various combinations of the 92 atoms that occur in nature. These are the 92 elements listed in the periodic table of elements that you may well have found so boring in your school chemistry class. Try to pay attention now though because this is important…

Atomic action

There are three basic bits to an atom, which are commonly referred to as sub-atomic particles—protons, neutrons and electrons. The protons and neutrons stick together to create the nucleus at the heart of the atom and the electrons whiz around the outside. It's like a tiny proton/neutron planet with orbiting electron moons. Different atoms have different numbers of protons, neutrons and electrons. For example, if 13 protons and 14 neutrons have formed a nucleus with 13 electrons in orbit, you have an aluminum atom. Gather a few million more together and you can form them into a can to hold beans or even into a bicycle. Aluminum is a stable atom; its constituent parts are balanced and happy together.

If you apply energy to an atom, the electrons orbit faster and the whole thing becomes very agitated. It expands as its electrons orbit further from the nucleus. An individual electron will be quickly pulled back to its original orbit and when it does so it sheds the energy previously input to it in the form of a punch of energy called a photon. Light photons are the ones we can see, but there are others. When you apply so much energy to the atom that the structure of the atom breaks down, the energy release can be quite phenomenal, take on many different forms and start a chain reaction that causes other atoms to break down, too. When their structure is changing, atoms don't just throw off light photons, they can emit alpha particles (two protons and two neutrons bound together), which leave the scene of the turmoil at a speed of around 10,000 miles (16,000 km) per second. They can also emit beta particles (which are basically unwanted electrons), neutron rays (which are streams of rejected neutrons) or gamma rays, which are bursts of pure energy.

Daily bombardment

Sounds like a pretty awesome weapon, huh? Well, our planet is being bombarded with this stuff every second of every day—and so are you and I. Cosmic rays, from the sun and other stars, some of which may have exploded millions of years ago, are made up mainly of protons zipping along at almost the speed of light and alpha particles, which are also no slouches. When they enter our atmosphere they crash into atoms in the atmosphere and their energy is dissipated, although secondary, much weaker rays still travel down to the Earth's surface. By the time they reach you walking along the street, you don't even feel a tickle as they hit you. If you were hit with alphas, betas, neutron or gamma rays with their full force, however, they could really mess you up.

To protect himself from the harmful rays of the Hawaiian sun, Elvis surrounded himself with hula-honeys.

Because these forms of radiation can affect other atoms, knocking off an electron here or there, they can affect the atoms that go to make up your body. The loss of an electron, as you can see from all of the above, can cause an atom to change its characteristics. This means that the atoms in each cell of your body can change, the cells will then either mutate or simply die off. Mutations and cancers (where the cells die) are, of course, one of the nastiest consequences of a nuclear explosion where the nuclear reaction produces just the sort of radiation rays we have been discussing.

The atoms that make up human skin tissue are quite good at absorbing most forms of light radiation but one that we all know to be harmful is ultraviolet light. Ultraviolet is a radiation form of a wavelength, of a strength just below that of X-rays. It can't travel through solid forms like X-rays, it can't even penetrate glass, but it does pack enough energy to burn exposed flesh—a deadly ray that people have relied on for years to make them look tanned and healthy! Too much exposure to ultraviolet, we now know, can cause skin cells to mutate and produce skin cancers—good enough reason to wear a hat and slap on the sunblock.

Copycat rays

Yes, Mother Nature has got those ray weapons licked and, in the ultimate irony, man has finally learned from Mother Nature in order to produce one of his most cynical bombs. The so-called neutron bomb relies on the fact that neutron radiation, which can be used to generate powerful gamma rays, can allow you to capture an enemy's assets intact. While a tactical explosion will be required to produce the radiation, relatively speaking it needn't be that big a bang and the neutron beams and gamma rays can then pass through buildings, fittings and fixtures, leaving them undamaged while killing (eventually) the people sheltering inside those buildings.

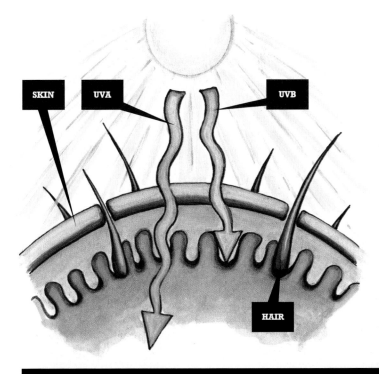

The two different wavelengths of ultraviolet light penetrate human skin to different depths. Both can cause skin cancer

Natural gas that is held under pressure underground is easily ignited, causing a major danger in oil exploration

A CASE OF GAS

THE WORST, MOST DESTRUCTIVE KIND OF EXPLOSION takes place in an enclosed space where the pressure of the blast fills the space and then has to force its way out. Perhaps the worst kind of enclosed space in which that could happen is a coal mine.

Even today, with modern technology and all of the precautions that are taken and safety checks that are made, mining is a dangerous business. Early last century, however, when factories, power stations, ships, steam trains and domestic heating relied on coal, there were far more miners working underground—without the aid of today's technology to keep them safe. In fact, they could be described as working inside a bomb where they themselves were the detonators.

At the coal face

In Britain between 1880 and 1910 there were over 1,000 deaths each year in the mines—an average of four deaths and 517 injuries every single day. In America between 1890 and 1907 there were more than 26,000 fatalities. Although rock falls were, and still are, the prime source of danger in a mine, the twin tragedies in America's coalfields in 1907 were caused by major underground explosions.

At the Darr Mine in Westmoreland County less than a week before Christmas in 1907, 239 men lost their lives. Miners tunneling 2 miles (3 km) inside the mine encountered a deadly pocket of methane gas. The gas, which had occurred naturally millions of years previously when the coal was first formed, had been locked in a cavity in the coal seam. Released from its pressurized containment, it streamed out to mix with the coal dust and oxygen and produce a highly explosive concoction.

Mine workers at that time used lamps with an open flame and the

Miners took canaries underground as the canary's respiratory system is so sensitive that it would begin to show signs of distress in the presence of gas or bad air

men's own lamps ignited the gas, producing a massive explosion and a searing flame that scorched through the tunnels. This tragedy came less than two weeks after America's worst-ever mining disaster in neighboring West Virginia. Around 380 men and boys had just begun their shift at the Monongah Mine near Fairmont when a shocking explosion rocked the entire area. Again, it was believed that gas, ignited by open lamps, caused the conflagration. The blast completely wrecked the mine workings, with an enormous cave-in trapping 360 miners underground. A 12-year-old boy worker was also killed at the surface of the mine.

In those days, children performed a variety of tasks underground and, although most countries introduced laws, it wasn't until 1938 that U.S. Federal regulation finally banned the employment of children in hazardous occupations. By 1941 mine workers' unions had agreed with employers that the minimum age for mine workers should be set at 18.

The world's worst-ever mine explosion probably occurred in Fushun in China where up to 3,000 may have died underground although the Chinese authorities have always been reluctant to provide exact figures.

Gas and oil

It is not only in coal seams that gases, accumulating since prehistoric times while trees and other organic life forms transformed over the years into coal, were first laid down. Natural gas is also trapped in the rock surrounding oilfields. Some of this is recovered and used to generate power or burned in our homes. Some of it is simply burned off at the well head in those huge leaping flames that you see high above the drilling platforms.

It is of course the pressure of this gas that can cause the oil to rush to the surface and spray all over the place like an erupting volcano when a drilling team strikes a deposit of oil.

A volatile mix

Working with the highly volatile mix of gas and oil, or gas and coal is never going to be anything but dangerous, although nowadays in coal mines safety checks are carried out to test for gas and to try to predict where pockets of gas might occur. Of course, there are no naked flames down mines nowadays, with electric lighting having replaced candle or oil lamps many years ago. In fact, injuries caused by electric shocks are now among the most common in mining.

BANG!

Pumping oil out of the ground can mean disturbing gas deposits that have been building up
since prehistoric times

The B-17 "Flying Fortress" was one of the foremost bomber aircraft of World War II

BOMBS AWAY!

SO WHO DO YOU THINK probably invented the world's first bomb? Well, I think we have to go back to the Chinese cook again (see page 59). The pot of black powder that went bang in his kitchen was probably the world's first man-made bomb. OK, so we have no way of knowing whether the Chinese cook ever really existed, but it's a safe bet that, having seen the destruction wrought by the black powder when enough of it went bang, the Chinese were quick to plant it in pots at the gates of enemy fortresses and hurl it at enemies on the battlefield. Indeed the Chinese are known to have been using a hand-held bomb, or hand grenade, almost a thousand years ago.

Hand grenades

When Europeans began experimenting with gunpowder and artillery, they realized that, at close range, men could hurl bombs just as well as a cannon. In the 15th and 16th centuries, small balls filled with powder and detonated by a lit fuse were being flung around the battlefield. These new bombs, hand grenades, worked best when thrown by the tallest, strongest soldiers and units of these giants became known as grenadiers. Hand grenades, although sometimes shunned because their unreliability could make them as dangerous to the grenadier as his enemy, have been in use ever since. They still provide a powerful, portable bang that can cause some very nasty damage to anything or anyone in close proximity when they go off.

Today, the most easily recognizable type of grenade is the time-delay grenade where removal of the safety pin releases a spring-loaded striker, which hits a percussion cap producing a spark. This spark ignites a fuse, that can burn for several seconds before igniting a charge, which then provides the energy to detonate the explosive packed into the grenade.

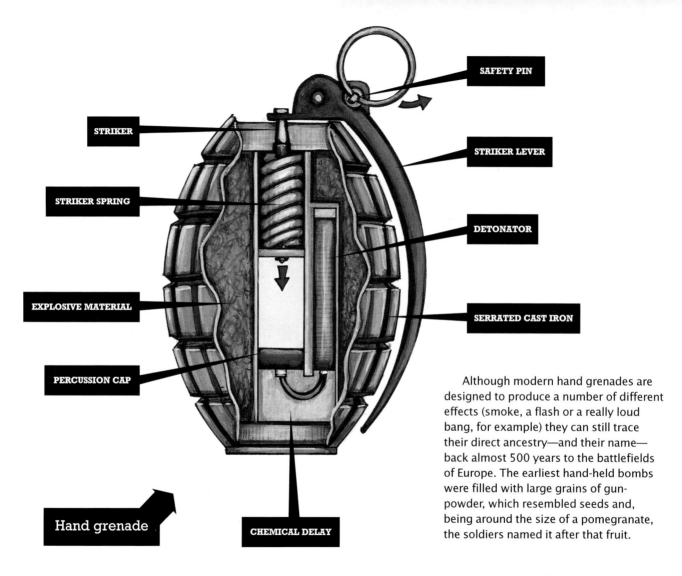

SAFETY PIN

STRIKER

STRIKER LEVER

STRIKER SPRING

DETONATOR

EXPLOSIVE MATERIAL

SERRATED CAST IRON

PERCUSSION CAP

Hand grenade

CHEMICAL DELAY

Although modern hand grenades are designed to produce a number of different effects (smoke, a flash or a really loud bang, for example) they can still trace their direct ancestry—and their name—back almost 500 years to the battlefields of Europe. The earliest hand-held bombs were filled with large grains of gunpowder, which resembled seeds and, being around the size of a pomegranate, the soldiers named it after that fruit.

A deck crew handling part of a modern aircraft's bomb load aboard an aircraft carrier

Early aerial bombing

At the other end of the scale when it comes to bombs are the biggest bombs ever developed. Although naval guns on the battleships of World War II could hurl a shell weighing as much as a small car a distance of over 20 miles (32 km), we don't really think of artillery rounds as bombs, so let's not count them. When we think of bombs what we really think of are those big black things that drop out of airplanes.

The first bomber aircraft started to appear during World War I when airplanes were first used in battle. At first it was thought rather unsporting to drop bombs from aircraft and there was a genuine reluctance among some military commanders to adopt the practice. It soon caught on, though. At first the pilots of rickety biplanes simply took a look to see what was below and then threw a bomb out of the cockpit in the general direction of the target. Needless to say, this wasn't very accurate!

Under the right conditions, an airship could provide a far better platform from which to aim your bombs and these giant balloons could carry a much bigger bomb a far greater distance than the airplanes of the time. German zeppelin aircraft even bombed London, although this became a highly risky business for the zeppelin crews. The giant cigar-shaped zeppelin balloons were filled with hydrogen gas which, being lighter than air, floated on air, lifting the whole airship into the sky. At over 600 feet (183 m) long, however, they were pretty big targets both for gunners on the ground and for pilots of the newly formed British Royal Flying Corps. The airships were so big, in fact, that RFC pilots could fly above them and drop bombs on them—bombing the bombers! The pilots had to beat a fairly hasty retreat if they scored a hit, though, because a balloon full of volatile hydrogen gas under pressure is little more than a bomb-in-a-bag! Add a flash to ignite it and the whole thing will go up.

The latest satellite guided bombs like the American JDAM weapon are claimed to be accurate to within a few feet, but who'd want to check?

Later developments

Throughout the course of the rest of the last century, a great deal of time, effort and expense went into the art of delivering bombs to a target from an aircraft. During World War II bombers were developed that could fly several miles high, in an attempt to avoid anti-aircraft artillery and enemy fighter planes. Sophisticated bomb sights were developed to help them deliver their payload on target, but accuracy was not always what it should have been. Needless to say, when the bombers missed their military or industrial targets, it was the civilians who suffered. In the end, the civilian populations of the combatant nations came to be regarded as legitimate targets in any case.

Today the waste of life, the unnecessary destruction and the waste of expensive munitions involved in poorly targeted bombing is deemed unacceptable by the military. For effective results the most highly accurate weaponry is most desirable. Accuracy means that today's conventional, non-nuclear bombs need not be as massive as the 22,000 lb (10,000 kg) Grand Slam bombs dropped by the RAF in World War II, although there are plans for a 30,000 lb (13,600 kg) bomb, which could penetrate to depths of up to 100 feet (30 m) below the surface before exploding.

The world's most accurate bomb is claimed to be the American JDAM (Joint Direct Attack Munition), which is continually monitored in flight by seven satellites to help guide it to within 6 feet (2 m) of its intended target. Guided missiles, of course, can be "flown" to their targets with even greater accuracy but the JDAM is the most effective "free-fall" bomb.

You certainly wouldn't want to be waiting on the ground with a measuring tape to check that, though, would you?

The first bomber aircraft were really only fighter or observation planes from which the pilot lobbed a couple of bombs at the enemy.

Zeppelin airships were used to drop bombs on London during WWI but were so big that British planes could fly above them and drop bombs on the bombers.

The grenade takes its name from the pomegranate fruit.

The Chinese were using a type of hand grenade as long as a thousand years ago.

SHAKE, RATTLE 'N' ROLL

WE HAVE ALL SEEN PICTURES of His Holiness the Pope, bending down to kiss the tarmac whenever his plane arrives at an airport. He is obviously thankful to be back on *terra firma*—solid earth. But is the earth really that solid? It certainly feels like it if you slip on a banana skin and plant your tail bone on the sidewalk! On the other hand, you are not really heavy enough to make much impact on the ground. How about that bulldozer rumbling by, though? Some big earth movers really do feel like they're moving the earth but to tell the truth, a bulldozer, bus, jumbo truck or even a tank isn't going to make the earth quake that much more than you do when you plant your backside on the sidewalk. But it will move a bit. Heavy impacts send shock wave vibrations through the surface of the earth like ripples on a pond. Strong enough vibrations will cause what we know as an earthquake.

Plates and fault lines

Earthquakes can be caused by underground explosions, heavy-duty impacts like a meteor strike, landslides or volcanic eruptions but most of them are associated with movements in the surface of the earth itself. It was discovered last century that the surface of the earth is not one continual crust. It is, in fact, seven large bits of crust called plates and a number of smaller plates. What's more, these plates are constantly on the move but, a bit like your dad in his favorite armchair in front of the TV, they don't move much, or very fast! Over the life of the entire planet, though, the tiny movements each year add up. Take a look at the map of the world. It's no accident that South America looks almost the right shape to fit into the west coast of Africa. Once upon a time it did just that.

As these plates move, it stands to reason that they must be pulling apart from one another, pushing against

A major earthquake can, quite literally, shake a
building to pieces

one another or generally scraping about
a bit. When they pull apart, magma from
beneath the crust can be released to cool
and seal the gap (see page 102), but a
trench or ridge is likely to form none the
less. When they push against each other,
one may start to slide under the other,
again forming a sort of ridge, or they
might pile up against each other to form a
mountain range. Scraping about a bit can
cause all sorts of damage to the edges of
the plates. Try laying some playing cards
down on a table and mixing them around
to see the sort of thing I mean.

The place where two plates meet is
called a fault. The most famous fault line
in the world is probably the San Andreas
fault line, which marks the boundary of
the Pacific and North American continental
plates. It runs along over 650 miles
(1,050 km) of America's west coast,
mainly through California, and movement
of the two plates has caused some
spectacular earthquakes. Californians
are used to the odd rumble in the ground,
of course. And all of the rest of us really
should be, too. There are, after all, around
three million earthquakes around the
world every year—more than five every
minute of the day—although most of
them are too weak for us even to notice.

In some earthquake zones, buildings are made of lightweight materials and designed to collapse as safely as possible—this wasn't one of them

Shocking results

All the scraping, pushing and crunching that goes on with the earth's surface plates causes vibrations that send out shock waves, some of which can travel right through the planet, although on the other side of the world only the sensitive equipment used by seismologists (scientists who study earthquakes) will actually pick up the vibrations. Closer to the scene of the quake, however, the shock waves will cause major damage. One form of vibrating shock wave produced by the enormous forces involved in the movement of the earth's plates will actually cause the ground to "ripple" up and down. This can easily shake buildings apart. In Turkey in 1999 an earthquake in Izmit, an industrial town 50 miles (80 km) from Istanbul, destroyed 20,000 concrete buildings and killed an estimated 30,000 people.

Since this is a book about bangs, you've probably guessed by now that earthquake shock waves can also take to the air and these do create a rather loud, rumbling bang.

I doubt if that is at all what His Holiness expects to hear when he bends over on the tarmac, though.

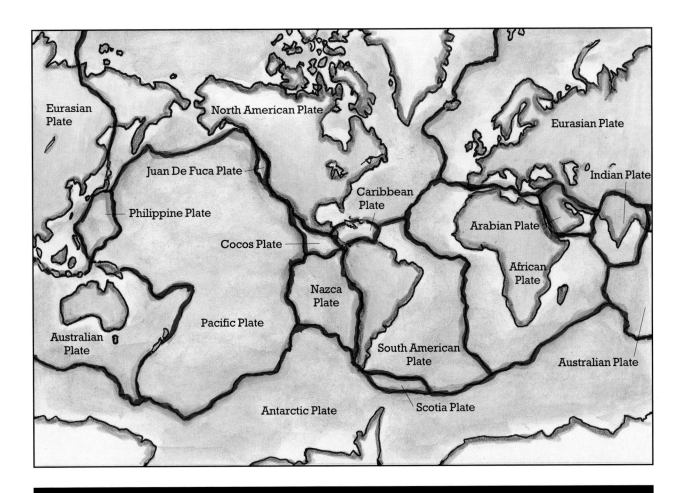

The surface of the earth is composed of seven major "plates" and a number of smaller ones

FAMOUS BANGS

THE HINDENBURG

LIGHTER-THAN-AIR AIRSHIP BALLOONS (also known as rigid dirigibles) were, prior to World War II, the last word in luxury air travel. Captain Max Pruss was commander of the biggest airship ever built, the *Hindenburg*. Entering service in 1936, the *Hindenburg* was the pride of Nazi Germany. She could carry 72 passengers all the way from Europe to the United States. Pampered by a crew of 61, the passengers could relax in the sumptuous lounge area of the cabin, enjoy gourmet food and listen to the pianist entertaining them on a baby grand piano.

The ticket price of $400 was expensive but it was the fastest way to cross the Atlantic—the Concorde of its day. The *Hindenburg* would cruise at an altitude of just 650 feet (200 m) at a speed of less than 80 miles (130 km) per hour, taking more than two days to complete the journey but conventional passenger aircraft flights across the Atlantic were not to exist for another three years. A two-day journey was still preferable to the eternity it took to travel between America and Europe by ship.

Airships are not too good in bad weather, as high winds tend to blow them about, so the *Hindenburg*'s transatlantic flights were limited to the summer months. The first flight of the 1937 schedule was May 3. The trip was becoming almost routine. The giant airship—more than 800 feet (245 m)

long—had crossed the Atlantic 10 times before, ferrying over 1,000 passengers to and fro. This trip, however, was to be far from routine.

Arriving slightly late over New York because of strong headwinds, Captain Pruss gave his passengers a quick glimpse of the Statue of Liberty before heading for Lakehurst Naval Air Station in New Jersey where the airship would be moored to a mast. As he approached the mast he adjusted the altitude of the airship by dumping 1,100 lb (500 kg) of water ballast, duly soaking the spectators below.

As the ship made its final approach to the mast a small flame appeared on the fabric of its hull near the tail. The airship achieved its lighter-than-air flight by being filled with hydrogen gas, a gas that is lighter than the air in our atmosphere. The airship balloon therefore "floats" in the sky. Unfortunately, hydrogen gas is highly inflammable and a spark on a hydrogen balloon of more than 7 million cubic feet (196,000 cubic meters) is the last thing you want. No one really knows where the flame on the *Hindenburg* came from but the explosion and subsequent fire consumed the airship in little over 30 seconds and it fell from a height of 300 feet (90 m). Amazingly, 62 of the 97 people on board survived the crash. Never before, or since, has a balloon gone bang in quite such a spectacular fashion.

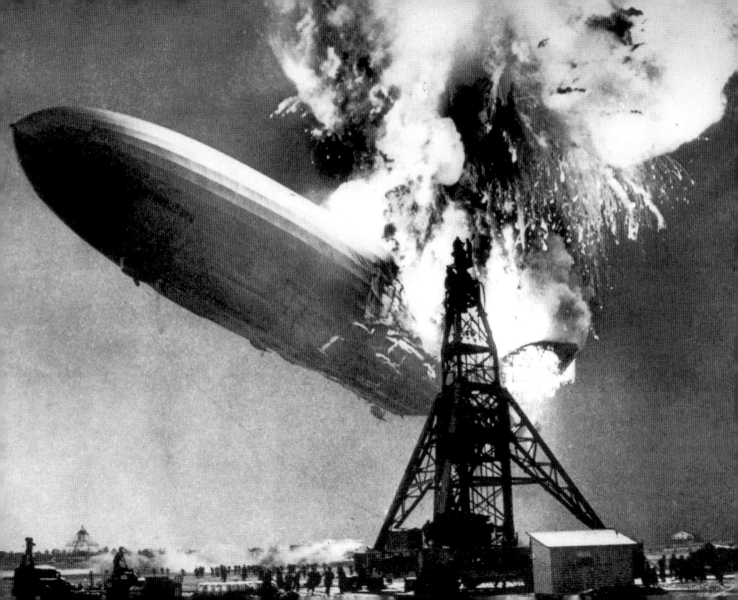

Lightning is one of nature's most spectacular manifestations of raw power

POWER STATION IN THE SKY

THUNDERSTORMS HAPPEN UNDER SPECIFIC CONDITIONS that tend to prevail at certain times of the year, so for most of us actually seeing lightning is not an everyday occurrence. But if lightning is not an everyday occurrence, then nobody has bothered to tell the lightning that, because it scythes through the sky every minute of every day all year—it's just not always happening in your neighborhood. In fact, there may be as many as 1,000 thunderstorms raging around the world every minute, producing about 10 flashes of lightning every second.

Accompanying the lightning is of course the thunderclap (see page 32), which can manifest as an incredibly loud bang if you are close enough to the center of the storm. The bang comes from a shock wave set up by the lightning bolt when it heats the air around it. Although most bolts of lightning may be about only 2 inches (5 cm) thick, they produce an intense heat of over 60 000°F (33 300°C), forcing the air to expand and create a shock wave heard as thunder.

So how is all this heat and the flash of light produced in the first place? What makes our sky such a wonderful power generator?

A flashy statement

First of all, picture how areas of warm air and cool air react to each other. Slightly warmer air, having been

If Benjamin Franklin had really flown a kite in a lightning storm to attract a lightning bolt, he would probably have been killed

heated by the sun or warmth rising off the surface of the earth, expands and is less dense than cooler air, so it will tend to "float" on the cooler air. The rapid rise and fall of air currents causes friction between the particles in the air and, in the water vapor of a cloud, cool water droplets or ice fall to the bottom of the cloud where the build-up of friction causes a negative electrical charge to accumulate. Conversely, a positive charge forms at the top of the cloud. The positive and negative areas of the cloud will regularly "short out," producing a huge electrical spark that most often occurs within the cloud or between one cloud and another.

When the negative charge at the bottom of the cloud is offered a way to discharge, it will take the easiest, most immediate option and a burst of lightning will be created. If this easy option is towards the ground, then a bolt of lightning will be seen streaking earthwards, but the shortest route to the ground is likely to be via the tallest thing in the area—a tall tree or the spire of a church. Rather than being one continuous flash, lightning is most often a series of incredibly short bursts that last only fractions of a second, which is why it can sometimes appear to flicker.

With a large thunderstorm producing as much energy as an atomic bomb, and up to a thousand of those bombs going off every minute, a vast amount of electrical energy is being created. If that sort of energy could be harnessed, much of the world's energy requirements could be easily fulfilled by Mother Nature—if only we could figure out how to fit the sky in a bottle!

At any given time, there are over 100 thunderstorms raging around the world, producing 10 flashes of lightning every second.

Most bolts of lightning are around 2 inches (5 cm) thick.

A lightning flash is more than five times hotter than the surface of the sun.

A 100-watt light bulb could be lit for around three months by the electric charge in a single flash of lightning.

ATOMIC EXPLOSIONS

NO ONE LIKES TO HAVE REALLY NOISY NEIGHBORS, do they? There's nothing worse than settling down for a quiet relaxing evening at home only to have the neighbors start setting off some of the loudest explosions you have ever heard. We may be the most highly developed creatures on Earth, but our neighbors—the thousands of other species that share our planet—must get pretty fed up when we start messing around making really big bangs. Of course, man hasn't yet produced anything in the big bang category that rivals what Mother Nature comes up with from time to time, but we have come pretty close with our nuclear efforts.

Nuclear explosions

The destructive force of a nuclear detonation is truly devastating and the aftermath can bring further horrors. This makes them effective as weapons—the threat of their deployment causes such terror as to add to their effectiveness—but how do nuclear bombs achieve this extraordinary destructive power?

To give you some idea of the amount of power involved, it is estimated that if all the energy from the detonation of 1 lb (0.5 kg) of TNT could be turned into heat, it would produce enough heat to boil about 4 gallons (18 liters) of water. If you could do the same with the energy released by 1 lb (0.5 kg) of uranium in a nuclear explosion, you could boil 25 million gallons (114 million liters) of water!

The energy is produced when an atom is either split (nuclear fission) or combined with another atom (nuclear fusion). The new atoms that are created are different in mass from those that existed before. With nuclear fission two new atoms might be created but they contain less matter than the original. With nuclear fusion, the new

The cloud of ash, dust and smoke from a nuclear blast rises as high as a commercial aircraft's cruising altitude

atom will contain less mass than the two original atoms. Where does the difference go? It is converted into energy. Just as a flame is produced as part of the chemical reaction when some forms of matter change state, energy is produced when atoms are split or combined. This is where that clever chap Einstein's famous equation $E=mc^2$ comes into play. The equation defines the way that when you convert matter into energy, the energy produced (E) is equal to the mass of the matter (m), multiplied by the speed of light squared (c^2). Given that the speed of light is 186,000 miles (299,350 km) per second, you're going to end up with a huge release of energy. That 1 lb (0.5 kg) of uranium we talked about earlier would be about the size of a large apple. The energy it produces is equal to that of around 1 million gallons (4.5 million liters) of gasoline—enough gas to fill every room on every floor of a five-story office block.

Fission and fusion bombs

In nuclear fission, the fissionable material that is used is naturally unstable, or radioactive, like uranium, making it easier to persuade it to decompose with an explosive input of energy. When the protons, neutrons and electrons of uranium atoms are bombarded with free neutrons, they soak them up. A uranium atom will accept a new neutron, become overloaded

ALBERT EINSTEIN

ASK A THOUSAND PEOPLE to name a famous scientist and 999 of them would probably say "Einstein." Albert Einstein was born in Württemberg, Germany, on March 14 1879. His father sold beds for a living and worked hard to provide an education for young Albert who showed great aptitude at school. Indeed, when the family moved to Milan in 1894, Albert stayed on at school in Munich to finish the school year. He then attempted to skip the rest of his high school education and go straight into the Eidgenössische Technische Hochschule (ETH) in Zürich to study for a degree in electrical engineering. Unfortunately, he failed the entrance exam, not scoring high enough marks in the arts section.

He successfully reapplied the following year and graduated from the ETH in 1900. During his time there he met his future first wife, Mileva Maric.

While working in the Swiss Patent Office, Albert developed revolutionary scientific theories including his special theory of relativity and, when still only 26, applied his theories to present the equation relating mass with energy, $E=mc^2$, with which he is so closely associated.

By 1911 Albert had become a professor at the University of Prague and the following year he moved to Zurich where he became Professor of Theoretical Physics at the ETH. Further academic honors followed and he became known as one of the greatest scientists of the century.

In 1932, while living and working in Germany, Einstein was identified as a Jew by the Nazis and moved to America. In 1939, he famously wrote to President Roosevelt warning him that the Germans were likely to try to develop an atomic bomb.

Albert Einstein died in America in 1955.

Einstein's famous equation defines how matter is converted into energy

$$E=mc^2$$

and divide into two new atoms. Between them the two new atoms will then expel two or three neutrons, which are soaked up by other uranium atoms in a sort of chain reaction. This reaction soon runs out of control and there is a massive explosion. In the earliest atomic bombs, this explosion happened before most of the fissionable

material could be used up. In fact, only 1.5 percent of it was used when the bomb named Little Boy was dropped on Hiroshima, Japan, in 1945. That made Little Boy only 1.5 percent efficient, yet it still produced a 14.5 kiloton explosion, equivalent to 14,500 tons (14,700 tonnes) of TNT. Fat Man, the name of the bomb dropped on Nagasaki three days after the Hiroshima explosion, was a slightly revised design and was 17 percent efficient, producing a 23 kiloton yield.

Nuclear fusion was found to produce even greater yields, with a fission bomb being used as the trigger to initiate a fusion reaction in a device that could produce an explosion almost a thousand times as powerful as the Little Boy bomb.

AFTERMATH OF A NUCLEAR **EXPLOSION**

THE DETONATION OF A NUCLEAR DEVICE PRODUCES A NUMBER OF EFFECTS. AROUND THE DETONATION SITE, THE HYPOCENTER, THERE WILL BE INTENSE HEAT AND RADIATING OUT FROM THE SITE WILL BE A SHOCK WAVE, JUST LIKE A NORMAL BOMB. SPLITTING OR FUSING ATOMS, HOWEVER, IMITATES THE ACTION OF THE SUN AND PRODUCES THE SAME KIND OF RADIATION THAT EMANATES FROM THE SUN.

Intense heat The temperature produced at the hypocenter is many times hotter than the surface of the sun and instantly vaporizes anything nearby. A ball of fire forms, extending several hundred feet and then rising into the air at around 100 feet (30 m) per second. The fireball consumes all of the oxygen in the vicinity, creating a vacuum that sucks in air and debris which, in turn, further feed the combustion. As the fireball rises it continues to expand until it is several hundred yards wide. The fireball is now a black, swirling ball of debris, soot, smoke and flame, which forms the mushroom shape associated with nuclear blasts as it approaches 25,000 feet (7,600 m).

Radiation On the ground, the explosion produces an intense flash, the visible light from this flash being accompanied by ultraviolet radiation as well as gamma rays. Although heat and light (including ultraviolet in small doses) are produced in a conventional explosion, in a nuclear blast they are hugely more powerful. The ultraviolet flash can cause the temperature of exposed skin to rise by 90°F (50°C) at a distance of over 2 miles (3 km) and flash burns will still be a problem over even greater distances.

Unlike a conventional explosion, the nuclear reaction also produces gamma rays. Although their range is not as great as the light radiation from the flash, gamma rays still travel at the speed of light and their effect on humans and animals can be hugely traumatic. In disrupting the particles in our bodies, gamma rays can cause cells to mutate or deform. This can lead to nausea and vomiting, cataracts in the eyes, hair loss and a loss of blood cells. Longer term, there is also a chance that cancers such as leukemia will follow and infertility and birth defects may also occur.

High-pressure shock wave The shock wave from a nuclear explosion is of course massive and can travel over several miles. It differs from the sharp wave of high pressure produced by a conventional explosion, not only in its size but also in its duration. Observers have described the high-pressure shock wave as being like a strong gust of wind going by. Closer to the hypocenter, that strong gust of wind will level buildings, fell trees and destroy almost anything in its path.

When the dust settles, there are still more problems. A debris of radioactive material, some of it unspent "fuel" from the bomb, will be dispersed over a wide area. This can continue to emit gamma rays or streams of radioactive particles for years, perhaps even centuries, after the blast.

The site of a nuclear explosion, or a nuclear test, therefore becomes a "no-go" area because of radiation after the blast, as does the site of a nuclear accident like the disaster at the Chernobyl nuclear power plant in the Soviet Union in 1986. The cloud of radioactive dust that emerged from Chernobyl raised radiation levels far and wide as it was carried across international borders by the prevailing winds. The site itself was eventually entombed in a concrete "sarcophagus," which sealed all of the radioactive material inside and is supposed to be able to last for hundreds of years.

WATCH THE BIRDIE

"**WATCH THE BIRDIE?**" Did photographers ever really say that when they had a family posing primly on a sofa in front of their camera? If not, then they are sure to have said something similar. The point was to get the whole family—husband, wife, children and dog—looking towards the camera. OK, so "Watch the birdie" might immediately make you look out the window, but not if the photographer really did have a bird in a cage just behind him. If you looked at the bird, you looked at the camera… and that might well be something you would immediately regret!

Early flash photography

Nowadays we are all used to electronic strobe lights for photography in poor light. The most basic of modern cameras generally have some kind of in-built "flash." In the early days of photography, though, that kind of gadget was still a whole century away. To illuminate their subjects, 19th-century photographers used flash powder. If you have ever seen a movie where an early photo session is recreated, you will have seen the photographer behind his camera under a black shroud to make sure that no light reaches his film plates except through the lens. He will be holding aloft a T-shaped bar. Along the top of the bar is spread flash powder. This powder could consist of a number of materials, but gunpowder and magnesium would have been favorites. In effect, it was a firework without the casing. Add a spark and there was a bright flash, a bang and a thick cloud of smoke. The children would get such a fright that they burst into floods of tears, the terrified dog would take off around the room barking his head off and everyone would start coughing from the effects of the smoke. In the instant before all that, hopefully, the photographer got his shot.

The reflector behind this photographer's flash bulb is as much for his protection as it is for lighting the subject

Advent of the flash bulb

Flash powder was eventually banned from the buildings of some organizations after a number of accidents (certainly no press officer nowadays would want a band of photographers wandering around, smoking pipes and carrying bags of explosive flash powder!) although it was the advent of the flash bulb that really saw it off. The flash bulb, however, didn't make that family portrait any the less traumatic. The very first flash bulbs contained magnesium in a glass bulb filled with pressurized oxygen... A gas under pressure with an explosive element that produces expanding gas when ignited, all encased in glass—the family would have been lucky to get out alive! To be fair, the earliest flash bulbs were designed for specialized photography, not family snaps, but later bulbs were equally hazardous, regularly spraying glass over the photographer's subjects.

Even in the 1930s, by which time flash bulbs using an electric current like ordinary light bulbs were widely available commercially, it was recommended that photographers used a silver, saucer-like reflector behind the bulb not only to reflect the light towards the subject but also to protect the photographer from flying glass!

Watch the birdie? You'd have to be cuckoo!

Although apparently solid and indestructible, ice is actually quite fragile

ICICLE POPS

"**WOULD YOU LIKE SOME ICE IN YOUR WATER, SIR?**" What a polite waiter—the only time you'd dream of refusing his request would be if you were sailing on the *Titanic*! When he's dropped a couple of cubes of the cold stuff into your glass, you hear it go *splosh—splosh—crackle*! What's that all about, then? That crackle is a series of pops or little bangs. Is it exploding ice? Not quite. The ice is being shattered, but it's not being blown apart by an explosion. It's being forced apart by itself.

Fatal flaws

When we think of a piece of ice, we tend to think of a cold, glassy cube as clear as water with smooth, slippery surfaces. Think again. Cold it may be, but is it really that clear? Unlikely. Like most diamonds and other precious stones, the ice cube has flaws—hairline cracks that can be either quite shallow or deep enough to stretch almost all the way through the cube. While the ice cube is in the freezer or in the ice bucket staying cold, these cracks stay closed. When dropped into your glass of water, the water penetrates the cracks. The overall effect of dropping ice into a glass of water is for the ice to lower the temperature of the water by drawing heat from it, thereby changing state and melting. The immediate effect, however, can be for the water that penetrates the cracks in the ice to freeze.

When water freezes, it expands. Ask anyone who's had frozen pipes during a wintry cold spell! The water in the pipe freezes and expands, cracking the pipe. When the weather gets warmer and the water thaws, it seeps out through the cracks and you discover that you have a leaky pipe. Much the same thing happens when the water seeps into the cracks in the ice cube. Suddenly surrounded by ice it instantly freezes and expands, forcing the cracks open and shattering the ice cube. The friction caused as the ice tears itself apart causes vibration, which transmits itself through the air to be heard by us as a crackling sound.

Apply the ice cube crackle to great mountains of ice weighing thousands of tons shearing off glaciers or the polar ice caps and the mini bang-crackle can become a very loud bang indeed.

A NIGHT AT THE THEATER

IT PROMISED TO BE A JOLLY EVENING. The play in performance at Ford's Theater in Washington DC was a comedy, *Our American Cousin*, and the president and first lady were looking forward to watching from the state box along with Major Henry Rathbone and his fiancée Clara Harris. Actor John Wilkes Booth was also looking forward to the night at the theater, but not because he was appearing in the play. He was planning to take on a much bigger, more historic role.

Assassination

The year was 1865 and the American Civil War had just ended with General Robert E Lee officially surrendering to Ulysses S Grant at Appomattox. Wilkes Booth was furious. A southern sympathizer and racist, he and a group of friends had been planning for almost a year to kidnap President Abraham Lincoln, hold him in the Confederate capital, Richmond, and demand the release of Confederate prisoners of war. The surrender put paid to the kidnap plan, but Lincoln's plans for the emancipation of the black population of the southern states were totally unacceptable to Wilkes Booth. Instead of kidnap, his plan became an assassination plot. Wilkes Booth's co-conspirators were to take the lives of Vice-President Andrew Johnson and Secretary of State William Seward on the same night at the same time, around 10.15 p.m. on Friday April 14.

President Lincoln arrived at the theater with his party at 8.30 p.m. and made his way upstairs to the state box. An hour later, Wilkes Booth arrived. A theater hand named Joseph Burroughs held Wilkes Booth's horse in an alleyway ready for his escape, and the assassin went into a saloon next door to the theater for a drink. Just after 10.00 p.m., Wilkes Booth returned to Ford's Theater. He approached the state box armed with a single-shot derringer pistol and a hunting knife. When he saw that the president's bodyguard was away from his post, Wilkes Booth opened the door to the box and fired at point blank range into the back of President Lincoln's head. That single bang assured Wilkes Booth of his place in history. Abraham Lincoln never regained consciousness and died nine hours later.

Retribution

Following a struggle with Major Rathbone, Wilkes Booth made his escape across the stage. He was later tracked down and shot while resisting arrest. Wilkes Booth's co-conspirators were all subsequently apprehended. Four were hanged and the others received lengthy prison sentences.

John Wilkes Booth claims his place in history by assassinating President Lincoln

HOT STUFF

Aaah... Doesn't that lovely warm sunshine bring a smile to your face on a warm summer's day? See how the flowers open and the birds bask in its warmth. It is the bringer of life to our planet, yet it is the most powerful ongoing nuclear explosion imaginable. In fact, it's probably not even imaginable, it's so awesome. It's the biggest bang happening in our galactic neighborhood... but it's not at all unique.

Our sun

Our sun is nearly 93 million miles (150 million km) away from Earth, far enough for light generated there to take over eight minutes to reach us and light, as we know, is a pretty fast mover. The distance, though, is just about right for us. Any closer and the sun would burn us to a crisp, any further away and things could get mighty chilly.

It's far from chilly on the "surface" of the sun. The sun is a burning ball of gas—70 percent hydrogen and 30 percent helium. To give you some idea of the size of the sun, the distance from where you are now sitting to the other side of the Earth—if you were to go in a straight line right through the center of the planet—is around 8,000 miles (12,900 km). If you were standing on the surface of the sun—where you would actually be hopping because your feet would be so hot!—the equivalent distance is 865,000 miles (1,392,000 km). Given that the surface temperature of the sun is about 9900º F (5500ºC), you'd need pretty thick soles on your shoes! The generator creating all this heat is at the sun's core where nuclear fusion reactions turn hydrogen atoms into helium atoms at a cozy 27 million˚F (15 million˚C).

Already you can tell that this is no ordinary gas cloud, can't you? If you could take a look inside the sun, you wouldn't even recognize it as gas. It's under so much pressure that it is more dense than lead. If the sun is part-hydrogen and part-helium, I hear you ask, and the nuclear fusion at its core is turning the hydrogen to helium, what happens when there's no more hydrogen? Good question. The sun has been doing its thing for around 4.6 billion years and eventually it will exhaust itself. When that happens there will be a really big bang and the sun will expand in size until it reaches Mars, pretty much

The sun

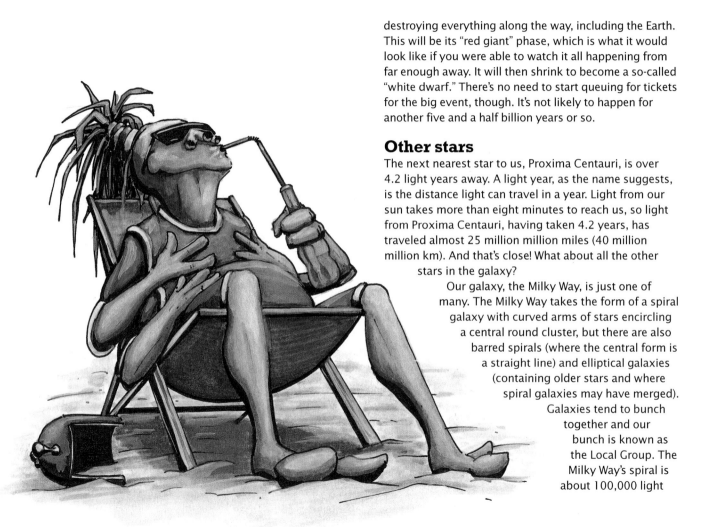

destroying everything along the way, including the Earth. This will be its "red giant" phase, which is what it would look like if you were able to watch it all happening from far enough away. It will then shrink to become a so-called "white dwarf." There's no need to start queuing for tickets for the big event, though. It's not likely to happen for another five and a half billion years or so.

Other stars

The next nearest star to us, Proxima Centauri, is over 4.2 light years away. A light year, as the name suggests, is the distance light can travel in a year. Light from our sun takes more than eight minutes to reach us, so light from Proxima Centauri, having taken 4.2 years, has traveled almost 25 million million miles (40 million million km). And that's close! What about all the other stars in the galaxy?

Our galaxy, the Milky Way, is just one of many. The Milky Way takes the form of a spiral galaxy with curved arms of stars encircling a central round cluster, but there are also barred spirals (where the central form is a straight line) and elliptical galaxies (containing older stars and where spiral galaxies may have merged). Galaxies tend to bunch together and our bunch is known as the Local Group. The Milky Way's spiral is about 100,000 light years across (that's more miles than I can be bothered to work out!) and our sun is about 25,000 light years from the center.

Have you ever tried to count the stars at night? Some of them are actually planets, some of them man-made satellites, but even if you could see all of the stars in our Milky Way galaxy it would take you a lifetime to count them. There are 100 billion stars in our galaxy. OK, you think you can count quickly? See how long it takes you to count to 100 and then multiply that by a billion. I'll sit here enjoying the sunshine until you've finished...

Space superbangs

The point is that the number of stars out in space is mind-numbingly huge. Although there are different kinds of stars they are all basically producing mega-explosions, superbangs in space just like our sun and, because it's happening in space, they are all producing silent bangs! With so many stars, it is reasonable to assume that there are millions upon millions just like our sun, each orbited by its own planetary systems. If this is the case, then there is a fairly strong chance that there are planets like ours orbiting suns like ours somewhere in a galaxy not too far away.

And the obvious conclusion? Life almost certainly does exist elsewhere in the universe. There are other creatures on other planets enjoying the warmth of other suns, just as we do.

LETHAL BANGS

THERE IS NO DOUBT that the most lethal bangs ever heard have come from small arms and hand guns of all descriptions. Early guns were hand built by craftsmen who turned gun-making into an art form. More recently, guns have been glamorized in movies and on TV, but it should be remembered that all small arms are designed for one thing—to kill. Moreover, most are designed specifically to kill humans.

Early guns

The earliest 13th-century "hand cannons" are barely recognizable as the forerunners of the guns we know today and in those days bows and arrows were still a far more reliable weapon with which to equip an army. By the early 14th century, however, this was beginning to change. Hand guns were still far from accurate, but a mass of gun barrels aimed at a mass of men could have a devastating effect as the metal ball ammunition could pierce most kinds of armor. Certain kinds of bow and arrow combinations were also capable of piercing armor, but a volley of gunfire was fast becoming a viable alternative, with the added advantage that the roar of the guns filled the enemy with fear.

The earliest guns that we would recognize as such today were matchlocks and flintlocks. The matchlock still worked pretty much like a shoulder-held cannon, whereby a glowing wick was applied to the gunpowder charge so as to shoot the ball out of the gun's barrel. The matchlock was eventually superseded by the wheel lock and then by the flintlock, guns which employed a trigger to release a wheel, and subsequently a hammer, which produced a spark when it struck the firing plate. The spark ignited the powder to discharge the weapon, now referred to as a "musket."

The problems with these early guns were that they were slow to reload because the powder and bullet had to be rammed down the gun barrel just as for a cannon, and in wet conditions damp in the powder meant that they sometimes didn't fire at all. With a bayonet stuck on the end, though, they were just as useful to the foot soldier as his old pike had been.

The advent of rifles

By the beginning of the 19th century, continued improvements and experimentation with firearms had led to the development of rifles. These guns had a groove spiraling up the inside of the barrel, which imparted a spin to the bullet. The spin helped the bullet to fly straight and made rifles far more accurate than the old muskets. They were still slow to load, though, giving them a poor rate of fire.

This problem was addressed in the mid-19th century when ammunition combining a percussion cap with a powder charge and bullet was developed. With a round of this new combined ammunition in the breech of the gun, a spring-loaded hammer or bolt was "cocked" by pulling it back until it locked into position. Pulling the trigger released the hammer or bolt, which then struck the percussion cap. The cap produced a spark, which ignited the powder charge, and the bullet was fired out of the barrel.

Naturally, there have been many further developments in the last 150 years, but this is the firing system still used in rifles and pistols today.

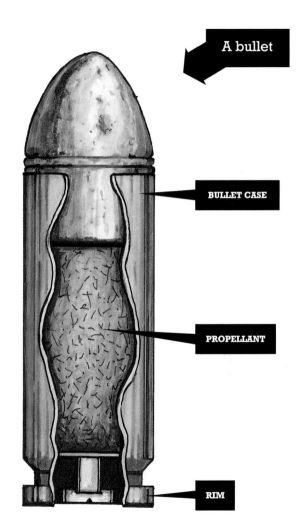

A bullet

BULLET CASE

PROPELLANT

RIM

Hiram S Maxim with one of his early machine gun designs

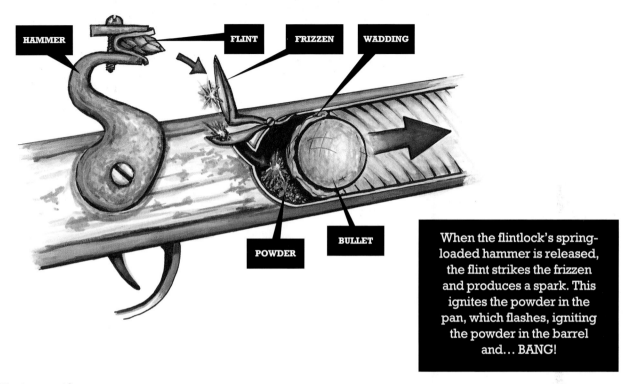

HAMMER

FLINT

FRIZZEN

WADDING

POWDER

BULLET

When the flintlock's spring-loaded hammer is released, the flint strikes the frizzen and produces a spark. This ignites the powder in the pan, which flashes, igniting the powder in the barrel and... BANG!

Automatic weaponry

Isaac Newton's third law of motion comes into play when a gun is fired. The expanding gases that force the bullet up the barrel of the gun push back on the gun at the same time—thus for every action there is an equal and opposite reaction.

This violent "kick-back," or recoil, caused a nasty bruise to the shoulder of Hiram S Maxim when he was target shooting one day. He decided to put the effect to use. His idea was to use the recoil force to push the bolt or hammer back into its cocked position. The same movement would eject the spent cartridge case that had held the gunpowder charge and load another round into the breech of the gun. Although other rapid-fire guns had been developed in the past, notably Dr Richard Gatling's "Gatling Gun," Maxim had come up with the first truly automatic system—the first machine gun.

FIREWORKS

WITH ALL THE DEVASTATION AND SUFFERING that so many man-made and natural bangs can cause, it's heartening to realize that most people never experience them. The majority of the world's population are never caught up in bombing raids or coal mine explosions, volcanic eruptions or nuclear blasts. Many people, however, do enjoy the far more pleasant experience of a fireworks display. And whom do we have to thank for that? Oh, no—not the Chinese cook again!

Yes, the Chinese were the first to use fireworks as an integral part of all sorts of celebrations. Superstition rather than science played a large part in this as the loud bangs were regarded as ideal for frightening off evil spirits, so fireworks were used to bring good fortune at, for example, weddings and the beginning of a new year.

Remarkably, a thousand years later, fireworks are still used today in exactly the same way all over the world.

Celebratory occasions

The spread in the use of fireworks started, as we saw way back on page 60, when explorer Marco Polo brought what became known as gunpowder back from the Far East to Europe in the 13th century. Gunpowder was already in use by China's closer trading partners in the Indian subcontinent and the Middle East but its introduction to Europe caused an absolute sensation. Soon, no ceremonial occasion or state celebration was complete without an extravagant fireworks display.

The Italians led the way of course, thanks to Marco Polo having first brought the black powder home to Italy. Different kinds of firework were constantly being developed and heads of state used breathtaking displays to entertain their subjects. Charles V of the Holy Roman Empire celebrated victory in battle with powerful fireworks displays, while Tsar Peter the Great was famous for elaborate New Year displays. The ruler renowned for producing the most flamboyant fireworks parties was the French king, Louis XIV, who regularly lit up the gardens of his Palace of Versailles with wondrous pyrotechnics.

Fireworks had become such a part of royal occasions by the 18th century that composer George Handel, a favorite of the English royal family and head of the Royal Academy of Music in London, produced a symphony called *Music for the Royal Fireworks*.

A professional fireworks display is the most delightful use of black powder ever devised

Shock value

Fireworks displays where we "ooooooh…" and "aaaaaah…" at the spectacular explosions are not just about big bangs, although those big bangs certainly contribute to the "aaaaaah…" factor. They can, in fact, quite literally take your breath away. Not only do you hear the really big bangs at a display, you feel them, too. The shock wave of high-pressure air will grip your chest like giant, although mercifully short, bear hugs, encouraging a sharp intake of breath to increase the air pressure in your lungs and fight off the momentary "crushing" effect.

Colorful chemicals

The main part of the "aaaaaah…" factor, though, is the shape and color of the explosions. Years of experimentation have led to different techniques being developed. The casing of a rocket, for example, is packed with gunpowder as well as small parcels of powder with different additives that produce different colors when ignited. The rocket has a lift charge ignited by the fuse that will send it up into the air. Its payload of small parcels will ignite a few seconds later. The shape of the parcels can determine the shapes of the bursts and the color depends on the ingredients mixed with the black powder.

A handful of chemicals are used to produce the most common colors. Red is produced by strontium nitrate, orange by iron, green by chlorate or barium nitrate, yellow by sodium and blue by copper. White bursts can incorporate aluminum, magnesium or possibly titanium.

Next time we watch those fireworks on July 4, maybe we should say a little "thank you" to that Chinese cook!

English king, George I, celebrated the signing of an Austrian peace treaty by hiring German composer Handel to write *Music for the Royal Fireworks* to accompany a display staged by an Italian technician.

French King, Louis XIV, was famous for his spectacular fireworks parties at the Palace of Versailles.

Different chemicals mixed with black powder produce different colors in fireworks bursts.

Russian Tsar, Peter the Great, hosted lavish fireworks parties to celebrate New Year.

INDEX

A

absorption of light 27, 29
aerial bombs 119-21
    see also specific types eg JDAM
air pressure 52-4, 155
airships 18, 119-20, 121, 126-7
alpha particles 109
anvil, bone 91, 92
arrows & archery 82, 95, 148
artillery pieces & cannon 56-7, 82, 94-6, 119
    black powder use 59, 60, 81-2, 94-5
    internal combustion engine comparison 37-8
atomic bombs 8, 18, 25, 43, 45, 132-7
atoms 26, 109, 111
    reactions *see specific types eg* fission *or events*
    eg Big Bang
Auchinleck, Claude 56-7

B

balloons 52-4, 83
    see also airships
Barnes Wallis 44, 46, 101
barred spirals 146
beta particles 109
Big Bang 10-17, 18
black powder 18, 59-60, 81-2, 94-5, 117-18, 152-5
blasting & mining 60, 112-14
blue skies 29
bomber aircraft 119-21
    see also specific bombs used eg JDAM
bombs *see also specific types eg* atomic bombs
bones of the ear 91, 92
Booth, John Wilkes 142-3
bottle rockets 84-5
bouncing bombs 46, 101
Braun, Wernher von 21, 82
breech-loaders 96

bubbles 76-7, 101
bulbs, flash 139
bullets 149

C

C4 18, 43, 45, 63
cannon *see also* artillery pieces & cannon
car engines 37-9
carbohydrates 48-9
carbon dioxide 76-7
chain reactions
    atomic 109, 135
    explosives 41
    incendiary 49, 75
    sound & hearing 92
Champagne 76-7
chemical reactions 25-6, 40, 43, 45, 75, 133, 155
Chernobyl disaster 137
civil engineering uses 60, 63
cochlea 91, 92
colors 26, 27, 29, 31, 83, 107, 155
combustion (car engines) 38, 39
Commando Vault bombs 45
composition 4 (C4) 18, 43, 45, 63
compression 38, 39, 88-9
conduction 72-3
cones & rods 30, 31
Congreve, William 82
convection 73
core, Earth's 10, 102
cosmic rays 109
crust, Earth's 102, 103, 104, 122-5
custard bombs 48-51
cyclotrimethylenetrinitramine (C4) 18, 43, 45, 63
cylinders, car engines 37-9

D

Daisy Cutter bombs 45
Dam Busters 46, 101
debris & shrapnel 43, 67, 75
detonation & detonators 41, 63
Dresden firestorms 74, 75
drums 68-9
dust explosions 48-51, 58
dynamite 18, 62

E

ears, ear-drums & hearing 54, 69, 90-3
    see also sound production
Earth, composition & temperature 10, 102-3, 122-5
earthquake effect bombs 43-5
earthquakes 18, 122-5
Einstein, Albert 133, 134
El Alamein 56-7
electric storms 32, 128-31
electromagnetic radiation 107-11
    see also specific types eg infra-red
electrons 26, 109, 111, 133
elliptical galaxies 146
energy 71-2
    mass-energy equation 133, 134, 135
    see also specific types eg light
engines
    internal combustion 37-9
    jet 86-9
eruptions, volcanic 34-5, 102-5
exhaust, car engines 38, 39
expanding gases
    Champagne 76
    engines 38, 89
explosives 41-3, 60-1, 62
    guns 151
    rockets & cannon 82, 95

sound production 90-3
thunder & lightning 32
underwater 101
volcanoes 103
explosions, relationship with fire 25, 27, 40, 58
explosives 41-3, 58
    see also specific explosives eg TNT
eyes & sight 27, 28-31, 107

F

faults, Earth's crust 123
fermentation 76
fire 25-6, 40, 48-9, 58, 75
fireballs 67, 136
firestorms 74, 75
fireworks 18, 60, 152-5
fission, nuclear & atomic bombs 8, 18, 25, 43, 45, 132-7
fizzy drinks 77
flash bulbs 138-9
flintlocks 148, 151
flying bombs, V-1 & V-2 21, 47, 82
four-stroke combustion cycle 38-9
Franklin, Benjamin 130
freezing
    nitroglycerin 19, 62
    water 141
    fusion, nuclear 132, 135
    see also Sun

G

galaxies 14, 15, 16, 146-7
gamma rays 28, 109, 111, 136-7
gases
    convection in 73
    see also expanding gases
gas clouds formation 14
in magma 103, 104

see also specific gases eg helium
glass 28
grain dust explosions 48-9
Grand Slam bombs 44, 45, 47, 121
gravity 14, 67, 83
grenades 117-18, 121
ground shock waves 43
gunpowder see black powder
guns 142, 148-51
    see also artillery pieces & cannon

H

hair dryers 86, 88
Hamburg firestorms 74, 75
hammer, bone 91, 92
hand grenades 117-18, 121
hand guns & small arms 142, 148-51
Hawaiian eruptions 104
hearing & ears 54, 69, 90-3
    see also sound production
heat & heat transfer 29, 71-5
Heinkel, Ernst 87
helium 14, 15, 17, 144
high explosives 60-1
    see also specific explosives eg TNT
Hindenburg 18, 126-7
Hubble, Edwin 15, 16, 17
hydrogen 14, 15, 17, 18, 120, 126, 144

I

ice & ice cubes 140-1
impacts, meteorite 22-3
incendiary bombs 75
infra-red 18, 28
intake, car engines 38, 39
interference, light 27
internal combustion engines 37-9

J

JDAM (Joint Direct Attack Munition) 120, 121
jet engines 86-9
Jor-el 19

K

Krakatoa 34-5, 102
Krypton 19

L

lava 104
light 14, 19, 25-31, 83, 107, 133
light years 146
lightning & thunder 32, 128-31
Lincoln, Abraham 142-3
low explosives 59, 60
    see also black powder

M

machine guns 150, 151
magma 102, 103, 104, 123
mantle, Earth's 102, 103
Marco Polo 18, 60, 152, 154
mass-energy equation 133, 134, 135
matchlocks 148
Maxim, Hiram S. 150, 151
medical uses 47, 63
memory banks 31, 93
meteorite impacts 22-3
methane 113-14
mineral waters 77
mining & blasting 60, 112-14
Mont Blanc 19, 78-9
Montgomery, Bernard 57
munitions ships 19, 78-9
muskets 148-9
muzzle-loaders 96

N

natural gas & oil 58, 112-15
neutrons, neutron bombs & neutron rays 14, 109, 111, 133-5
Newton, Isaac 83
Newton's third law of motion 82, 89, 151
nitroglycerin 18, 19, 61-2, 63, 75
Nobel, Alfred 18, 62
nuclear reactions *see* fission; fusion; *see also specific events eg* Big Bang

O

Ohain, Hans von 86-7
oil & natural gas 58, 112-15
oval window 92
oxygen 19, 26, 67, 75, 89, 101, 136

P

particles, subatomic *see specific types eg* protons
party balloons 52-4, 82
photography, flash 138-9
photons 26, 27, 109
pinna 90, 91
pistons 38, 39
planets 14, 102, 147
plastic explosives 63
plates, Earth's crust 122-5
Plinian eruptions 104
Polo, Marco 18, 60, 152, 154
pressure 52-4, 61, 100-1, 103-4, 155
protons 14, 109, 133
pyrotechnics 19
    see also fireworks

Q

quarks 13, 14, 20

R

radiation, electromagnetic 107-11
    see also specific types eg infra-re
radiation, heat transfer 72, 73
rainbows 27
RDX (C4) 18, 43, 45, 63
reactions *see also specific types eg* chain reactions
recoil, guns 151
reflection & refraction, light 27, 28, 29
retina 29, 30, 31
rifles 149
rockets 20, 81-5, 89
    fireworks 155
    V-2 flying bombs 21, 47, 82
rods & cones 30, 31
Rommel, Erwin 56-7

S

San Andreas fault 123
scattering, light 28-9
scoria cone volcanoes 104
semicircular canals 91, 92
shaped charges 63
shield volcanoes 104
shock waves 41-7
    atomic bombs 43, 136, 137
    balloon bursts 53-4
    Champagne corks popping 76
    earthquakes 122, 124
    fireworks 155
    medical uses 47
    nitroglycerin 62
    sonic booms 99
    in space 67
    thunder & lightning 32, 129
    underwater 101
shrapnel & debris 43, 67, 75

sight & eyes 27, 28-31, 107
singularity 12, 16
sky color 29
small arms & hand guns 142, 148-51
Sobrero, Ascanio 19, 61-2
sodas 77
sonic booms 98-9
sound production
    balloon bursts 54
    drums 69
sonic booms 98-9
sound waves 90, 92, 93, 99
    underwater 100-1
    in a vacuum 14, 17, 64
    vocal cords 6, 92-3
    see also hearing & ears
space, explosions in 64-7
speed of light 26, 27, 133
spiral galaxies 146
stars 14, 102, 109, 146-7
    see also Sun
stirrup, bone 91, 92
storms 32, 128-31
stratovolcanoes 104
Strombolian eruptions 104
subatomic particles *see specific types eg* protons
Sudbury meteorite 22-3
Sun 10, 20, 72, 107-11, 144, 144-7
supersonic flight 98

T
Tallboy bombs 44, 46, 47
telescopes 15
temperature
    atomic bombs 136
    Death Valley 10, 11
    Earth's core & mantle 10, 102

firestorms 75
Krakatoa effect 35
lightning bolts 129, 131
nitroglycerin explosions 75
nitroglycerin freezing point 19, 62
Sun 10, 144
universe at Big Bang 11, 12, 13, 14, 16, 17
terrorist bombs 43, 45, 63
thunder & lightning 32, 128-31
time-delay grenades 117-18
TNT (trinitrotoluene) 20, 62-3, 78, 133
toluene 62-3
tsunami 35
tympanic membranes (ear-drums) 54, 69, 90-2

U
ultraviolet 20, 28, 111, 136
underwater explosions 100-1
universe, creation & expansion 10-17

V
V-1 & V-2 flying bombs 21, 47, 82
vacuums
    atomic bomb effects 136
    sound in 14, 17, 64
vacuum cleaners 86, 88
vacuum effect 45-7
vibrations
    bangs 90, 93
    drums 69
    ear-drums 54, 69, 90-2
    earthquakes 122, 124
    ice crackling 141
vocal cords 6, 92-3
vision & eyes 27, 28-31, 107
volcanoes 34-5, 102-5
von Braun, Wernher 21, 82

von Ohain, Hans 86-7

W
Wallis, Barnes Neville 44, 46, 101
Warsitz, Erich 87
water pressure 100-1
waves
    light 27
    sound 90, 92, 93, 99
    tsunami 35
    see also shock waves
wheel locks 148
white light 27, 83, 107
Whittle, Frank 86-7
Wilkes Booth, John 142-3

X
X-rays 21, 28, 107, 111

Y
Yaeger, Chuck (Charles Elwood) 21, 98, 99

Z
Z particles 21
zeppelins, airships 18, 119-20, 121, 126-7

PICTURE CREDITS

Corbis: 79,87,110,139,150
Essential Works: 19, 20, 24, 26, 114, 128 and back jacket.
Getty Images: 70
Hale Observatories/Science Photo Library: 15
Rex: 62,116
Topham Picturepoint: 11,34, 36,41,46,55,56,59,69,80,83,97, 98,103,106,112,119,123,124,127,133,134, 140,143,153 and back jacket.

ACKNOWLEDGMENTS

With huge thanks to the team at Essential Works, 168a Camden Street, London NW1 9PT, England

Design: Barbara Saulini
Editorial: Dea Brøvig and Jo Lethaby
Illustration: Neil Jones
Picture Research: Nikki Lloyd
Office Dogsbody: Nellie